Microfabrication for Industrial Applications

Micro & Nano Technologies Series

Published Titles

9780815515739	Francois Leonard	*Physics of Carbon Nanotube Devices* (2009)
9780815515784	Mamadou Diallo, Jeremiah Duncan, Nora Savage, Anita Street & Richard Sustich	*Nanotechnology Applications for Clean Water* (2009)
9780815515876	Rolf Wüthrich	*Micromachining Using Electrochemical Discharge Phenomenon* (2009)
9780815515791	Matthias Worgull	*Hot Embossing* (2009)
9780815515838	Waqar Ahmed & M.J. Jackson	*Emerging Nanotechnologies for Manufacturing* (2009)
9780080964546	Richard Leach	*Fundamental Principles of Engineering Nanometrology* (2009)
9780815520238	Jeremy Ramsden	*Applied Nanotechnology* (2009)
9780815515944	Veikko Lindroos, Markku Tilli, Ari Lehto & Teruaki Motooka	*Handbook of Silicon Based MEMS Materials and Technologies* (2010)
9780815515869	Matthew Hull & Diana Bowman	*Nanotechnology Environmental Health and Safety* (2009)
9781437778236	Yugang Sun & John Rogers	*Semiconductor Nanomaterials for Flexible Technologies* (2010)
9781437778403	Victor Zhirnov & Ralph K. Cavin	*Microsystems for Bioelectronics* (2010)
9781437778489	Zeev Zalevsky & Ibrahim Abdulhalim	*Integrated Nanophotonic Devices* (2010)
9780815515456	Yi Qin	*Micromanufacturing Engineering and Technology* (2010)
9780080964478	Jeremy Ramsden	*Nanotechnology: An Introduction* (2011)
9780815515821	Regina Luttge	*Microfabrication for Industrial Applications* (2011)

Forthcoming Titles

9781437778632	Gurumurthy Ramachandran	*Assessing Nanoparticle Risks to Human Health* (2011)
9781455778621	Karthikeyan Subramani & Waqar Ahmed	*Emerging Nanotechnologies in Dentistry* (2012)

Microfabrication for Industrial Applications

Regina Luttge

ELSEVIER

AMSTERDAM • BOSTON • HEIDELBERG • LONDON
NEW YORK • OXFORD • PARIS • SAN DIEGO
SAN FRANCISCO • SINGAPORE • SYDNEY • TOKYO
William Andrew is an imprint of Elsevier

William Andrew is an imprint of Elsevier
The Boulevard, Langford Lane, Kidlington, Oxford, OX5 1GB, UK
225 Wyman Street, Waltham, MA 02451, USA

First published 2011

Notices

Knowledge and best practice in this field are constantly changing. As new research and experience broaden our understanding, changes in research methods, professional practices, or medical treatment may become necessary.

Practitioners and researchers must always rely on their own experience and knowledge in evaluating and using any information, methods, compounds, or experiments described herein. In using such information or methods they should be mindful of their own safety and the safety of others, including parties for whom they have a professional responsibility.

To the fullest extent of the law, neither the Publisher nor the authors, contributors, or editors, assume any liability for any injury and/or damage to persons or property as a matter of products liability, negligence or otherwise, or from any use or operation of any methods, products, instructions, or ideas contained in the material herein.

British Library Cataloguing in Publication Data
A catalog record for this book is available from the British Library

Library of Congress Control Number: 2011926754

ISBN: 978-0-8155-1582-1

For information on all William Andrew publications
visit our website at *www.elsevierdirect.com*

Typeset by: diacriTech, India

Printed and bound in Great Britain
11 12 13 14   10 9 8 7 6 5 4 3 2 1

Contents

Preface

Small, smaller, *nano*: "What happened?". Posing this question insinuates that "micro" has not made its way into the popular scenarios of opinion makers. Micro has had tremendous success in the microelectronics industry, but all other fields lack attention. It is not my current intention to evaluate why *micro* lacks impact on society and *nano* is overrated, but to provide a book on *Microfabrication for Industrial Applications*. This is because of two reasons: (1) I have spent the last twenty years in this field of research and I am fascinated by the fact that not only can we fabricate a hole of $10\,\mu$m diameter, but also we can make it at a precision of $10 \pm 0.05\,\mu$m under mass production conditions, and, (2) I am not only a scientist in this field but also an entrepreneur.

Having spent a decade of my professional career in Twente, the Netherlands, it is inevitable that I should present some examples of its successful spin-off activities, specifically, from the University of Twente's MESA+ Institute for Nanotechnology.

Let me briefly take the opportunity here to give you some facts about the University of Twente. It was established at the very east of the Netherlands in 1960, with the purpose of upgrading the region's economy through education and employment. Today, it is one of the largest organizations in the region, with 3300 employees and 9000 students studying high-tech sciences, (medical) engineering technologies (including ICT), business and public administration, communication, and psychology. The University's slogan, "High-tech, human touch" is not just a phrase; it has become a center for cultural and economical redevelopment of an environment that was previously in decline.

I would also like to refer to some of the influences that helped me to produce the book in its present form. First of all, I have to thank the Dutch Science and Technology Foundation, STW, for funding my research activities, and the University of Twente and the Science Park of Twente for providing me with a working environment in which it was possible to dedicate my time and effort to the book. I was also able to learn that technology transfer is about people as much as technology.

The research group Mesoscale Chemical Systems is affiliated with the MESA+ Institute for Nanotechnology, headed by Prof. Dr. J.G.E. Gardeniers. It provides me with a superb facility for research on micro-nanofabrication.

I am therefore very much indebted to Prof. Dr. J.G.E. (Han) Gardeniers. He has given me a scientific home in his research group.

This book was intended to be joint effort between Han and myself, about three years ago. Due to the many tasks resting on Han's shoulders in his position as head of the group, he asked me to write it alone. When we suggested this solution to the publisher, Elsevier, I was overwhelmed with the trust that they all had in me, and I was determined to complete the project for the reasons stated above, and also because I read the second year undergraduate course Lab-on-a-Chip Technologies in a project-oriented approach. I felt that a book dedicated to microfabrication for industrial applications could bring together the endlessly scattered background information onto one line of thought, which would enable students to familiarize themselves with the design of future Lab-on-a-Chip products, as well as to get a general understanding of the needs for miniaturization in society. The present book, therefore, provides an introduction to the topics around the concept of the Lab-on-Chip and related fields of microfabrication technologies, for students and non-specialists.

Thanks to Matthew Deans, Senior Publisher, and Frank Hellwig, Editorial Project Manager at Elsevier; I realize that it is OK to restrict the content, instead of "wanting to write it all". I eventually made sufficient progress on the book project when I was ready to see that it made sense to reuse some of my previously produced texts. I refer, here, mainly to my own previously unpublished PhD thesis on "Fabrication technologies for optical scanners based on micromachined cantilevers", (Imperial College, London, 2003), which laid the foundations for Chapters 2 and 5, and three specific publications, which I (co)authored between 2007 and 2010, that gave an overview of nanolithographies for nanoarrays (discussed in Chapter 4) and the utility of microchip capillary electrophoresis (discussed in Chapter 7). I present these parts here in one coherent line of thought. I noticed that the text that was originally intended as a scientific contribution fulfils its purpose as a standalone document in the expert community, but not necessarily as a source for students or non-experts who wish to learn about micro-nanofabrication. Therefore, I felt that it was permissible to incorporate these previous texts as central core information in this book.

This process also made me realize that if students see a gap between scientific information, and that supplied for educational purposes, it may also be a hint for bridging what society experiences as the *technology gap* between science and its application. The objectives of this book are first to allow the reader to become familiar with the field of microfabrication, and second to give the reader a guideline for assessing the potential of any new technology that is based on microfabrication, thus reducing the investment risk by means of a simple academic tool: desk research.

This book is set out in nine chapters. Chapter 1 gives a brief introduction to the value-creation process of a commercial technology implementation chain by securing a patent. Chapters 2–4 illustrate the technical background knowledge of process engineering for micro-nanofabrication purposes. Chapter 5 concentrates on a more classical microtechnology field: the microtransducer, while Chapters 6 and 7 present an introduction to (bio)chemical sensors and microcomponents, and the technologies used in the life-sciences industry. Chapter 8 illustrates the concept of academic desk research, which is applied to evaluate microneedle array technology for its application in diabetes management. This specific application is far beyond the original scientific project goals. However, it is via this type of *project after care* analysis that I am able to begin screening the depth of societal and economical impact of the novel technologies developed in the lab. I propose that the latter allows scientists to get in touch with key-technology managers in relevant industries. Eventually, we all have to realize that it is industry which must balance the budgets of these identified values with the actual available resources for development throughout a product's life-cycle.

The book could not have been written without the support of others. A considerable part of my scientific research consists of collaboration with other scientists and technologists. This work is incorporated to a large extent in Chapters 4, 7 and 8. Specifically in the work on nanofabrication for biomaterials, my thanks are due to Prof Dr. J.A. Jansen and his research staff at the Department of Biomaterials, Radboud University Medical Centre, Nijmegen, The Netherlands, and to my recent PhD student here at MESA+, Maciej Domanski, who combined most spectacularly in the course of his project the capabilities of laser interference lithography (LIL) into a new nanofabrication technology for medical-grade bulk titanium implants. The utility of LIL as a nanolithographical technique and its subsequent application is discussed in more detail in Chapter 4. The other extensive collaboration project I was involved with is the work on microneedle array technology discussed in Chapter 8, as an example of desk research, which is one important driver for innovation. With respect to this collaboration, I wish to thank Prof. Dr. R.J. Scheper and Dr. T.D. de Gruijl at Free University Medical Center, Amsterdam, The Netherlands, for their honest and motivating belief in this novel technology for advanced cancer vaccination therapies, by means of which they have also inspired spin-off applications of University of Twente's patent-pending ceramic, nanoporous microneedle array technology for other *SMART*[1] products based on this platform technology. I truly hope to be able to secure the required funds and continue this collaborative study to validate the benefits of these types of nanoporous microdevices. Within this context of *science enabling innovation* I owe thanks also to the professionals with whom I am involved via the organization and teaching of the Entrepreneur Workshop for Physicists and Engineers, established by the Institute of Physics, UK, London, and with whom I have the most extraordinarily exciting exchange of ideas on the topic of entrepreneurship.

My thanks also include all my colleagues, technicians and fellow scientists, who consistently work (and have worked) with me towards the understanding and implementation of micro-nanofabrication. Furthermore, I would like to acknowledge the dedicated support of Jacqueline Emmerich-Nijenhuis, our group's secretarial assistant, and Stefan Schlautmann, our group's technical assistant, who always lend a helping hand.

Of course, there are also others who have made my life comfortable up until now, and I should not forget to express my gratitude to them. First of all my thanks go to my family, my two brothers and their families and of course my parents, who are all based in my home town, Vienenburg, Germany. They do not see or hear much of me since I left to find myself by building an academic career, although they are always in my heart, and without their love I could not live as happily as I do. Of course, there is more than science. I am particularly lucky that I have found in Michel a partner with whom I can share my scientific passion and my life, and without whose inspirational support and the tips and tricks he revealed to me from his own book project, I would not have been able to accomplish the book.

Finally, microfabrication is the essential core of microsystems technology (MST) that focuses on the miniaturization of engineering systems to accommodate the design specifications of small space, light weight and enhanced portability. At the onset of all miniaturization we mainly refer to the development of the integrated transistor, the workhorse device by means of which all the major new markets were created. The importance of MST lies, for a large part, in the economical and technical development of innovative systems in

---

[1]SMART: Specific, Measurable, Attainable, Relevant, Tangible

general, and particularly in the extended functionality which is made possible by this approach. An example of such miniaturized systems is the wide-scale utility of distributed transducer networks. The research field of microfabrication technology was established in industry approximately 50 years ago, and thus it is a relatively young business discipline with a high potential.

Leaving the question from above: small, smaller, *nano*: "What happened?" for you to answer, by starting your own journey around the world of micro-nanofabrication, I hope that this book will spark many new ideas for the implementation of microfabrication in industrial applications.

Last but not least: thank you. Your comments and suggestions for micro-nanofabrication implementations and its related topics are highly welcome.

Regina Luttge
March 2011, Enschede, The Netherlands
E-mail: r.luttge@utwente.nl

Author Biography

Regina Luttge was born and grew up in Germany, where she was awarded a BEng by the RheinMain University of Applied Sciences in 1993. She received an Erasmus award and spent half a year at the University of Central Lancashire, UK, where she was introduced to research studying liquid crystals. From 1994 to 1999 she had been working in Research & Development at the Institut für Mikrotechnik Mainz, Germany, first as a process engineer facilitating a variety of LIGA projects utilizing x-ray lithography and subsequently as a researcher on the application of electron-beam lithography for x-ray masks. She then undertook research in the field of microsystems technology at the University of London, Imperial College, UK, under the supervision of A.S. Holmes and was awarded a PhD in 2003 for her thesis 'Fabrication technologies for optical scanners based on micromachined cantilevers'. Since 2001 she has been a research member at MESA+ Institute for Nanotechnology, University of Twente, the Netherlands, where she conducted her postdoctoral research on Lab-on-a-Chip applications and nanolithography from 2003 to 2007. During her postdoctoral research she supported technology transfer and actively motivated the industrial utilization of research results. The latter led to her involvement in high-tech entrepreneurship and the initiation of her first own spin-off company MyLife Technologies in 2008, which is incubated by the business development program at University of Twente and the Valorization program of the Technology Foundation STW, applied science division of The Netherlands Organization for Scientific Research (NWO), and the technology program of the Dutch Ministry of Economic Affairs. MyLife Technologies received recently a first-phase Life Sciences Pre-Seed Grant by The Netherlands Organization for Health Research and Development (ZonMw). With respect to her academic career, Regina Luttge received a VENI award by NWO in 2007 and was appointed by University of Twente as assistant professor at the Faculty of Science and Technology. She currently holds a tenure-track position affiliated with the research chair Mesoscale Chemical Systems at MESA+ Institute for Nanotechnology. Her scientific research focuses on nanofabrication technology and the engineering of microsystems for applications in (bio)medical technology, with a specific interest in neuroscience. In her new research project she will combine nanotechnology and neuroscience for which she was awarded the ERC starting grant in 2011.

CHAPTER

Introduction

1

CHAPTER CONTENTS

1.1 PHILOSOPHY OF MICRO/NANOFABRICATION

Microsystems technology (MST) focuses on the miniaturization of engineering systems to accommodate design specifications of small space, light weight and enhanced portability. An additional advantage of such portable systems is their wide-scale utility in distributed transducer networks. The importance of MST lies, for a large part, in the economical and technical development of innovative systems that it makes possible. The field of microfabrication technology has been established for approximately 50 years and thus it is a relatively young discipline. The first object of miniaturization was the integrated transistor, the workhorse device by means of which major new markets were created. For example, information and communication technology (ICT) relies on the technical principles of miniaturization by integrating more and more electronic functional elements into the same restricted area of a silicon die, the chip. Complementing this chip with a large data storage capacity that has fast read/write access and a high-definition display has given rise to systems which have penetrated all layers of personal and professional human lives. These types of devices are a smart combination of millions of transistors on a single chip, produced on dedicated microelectronic production lines. Transferring technological innovation into a robust

Microfabrication for Industrial Applications. DOI: 10.1016/B978-0-8155-1582-1.00001-0

and efficient production and marketing process is the secret to providing more and more computing power. The accurate line-width control during manufacturing these devices, at the micrometer or subsequently the nanometer level, is the main reason for giving this pool of associated fields concerned with the design, fabrication, assembly and testing the names micro- and nanotechnology, respectively. The very nature of these new disciplines within the engineering sciences originate from the principles of miniaturization, which was based on having an increasing number of the same circuitized components available on one die. Eventually, further and further integration allows novel functions that were originally unforeseen by conventional machining techniques. One example of such a novel microfabricated product, combining optical and electrical functions with a mechanical function at a small footprint, is an integrated optical light modulator, which we simply call a *beamer* in everyday life and which we consider a commodity today.

The technical methods which are used to manufacture micro products are described collectively as microfabrication techniques. Their efficiency is due to the definition of patterns in a masking layer and the subsequent parallel transfer of these accurate patterns into a functional material. During pattern transfer, the pattern is therefore copied from a mask, which carries the design features, onto or into a work piece. This work piece can be defined within a thin film or a stack of films, or the bulk of the material, which is also used as the handling platform during such a sequence of process steps. These two distinct approaches are called surface and bulk micromachining. Usually, pattern transfer is a process of at least two main steps. The first step involves the generation of the pattern, either directly by a serial write process, or by parallel patterning using an exposure through a master into a layer (usually a photosensitive polymer). This layer then acts as a masking layer for the second main step: shaping the functional material by processes such as deposition, etching or implantation. Specialists in the field generally call this layer a resist. Other modifications of the accessible areas of the work piece are of course also possible.

The continuous implementation and definition of novel fabrication techniques for modifying and transferring the patterns is key to innovation in the market place. It is this field of research and development that I would like to address in this book.

Lithographic processes have been described many times. In brief, the careful definition of process steps brought about the initial success of industrial lithography, specifically photolithography utilizing a mercury lamp with an intensity peak at a wavelength of 365 nm (UV-light) for transferring the information contained in the master into the resist. An average process sequence for a single device may contain 20–30 individual steps, some of which are considered to be the main steps of the process. This main step receives greater attention during the development of a process document. However, any set or combination of these steps may lead to a new technology.

MST is considered as less conservative than integrated circuit technology (microelectronics), because MST processes are reshaped by prototyping or the fabrication of demonstration devices instead of focussing on high-volume production. For microelectronics it is important that each process step is optimized for through-put and robustness (achieving a high yield). Costs are reduced in microelectronics manufacture due to the ability to copy a complex design in parallel from a special information carrier (photomask) hundreds and thousands of times, with the same high pattern fidelity, by simple shadow optics (exposure through a mask). This manufacturing strategy is called batch processing and is also an essential aspect of the philosophy of micro- and nanofabrication. An overview of the procedure which establishes a societal need, develops the technology and realizes an MST device is depicted in Figure 1.1.

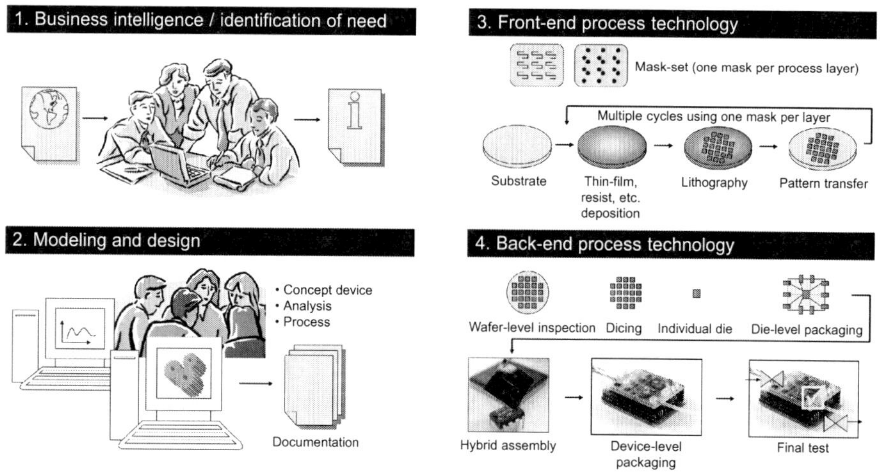

FIGURE 1.1

Schematic of the process from the indentification of a need to the development of batch-process technology.

Nanofabrication is a logical step to the further downscaling of the physical size of components and functional elements, often using the same machinery as microfabrication. Nanotechnology as a new discipline, however, should not be considered as simply an extension of existing techniques to smaller dimensions, but as the integration of novel functions based on an understanding of interactions at a scale of less than 100 nm. Alongside this, *bottom-up* manufacturing has been introduced. This approach or the previously used *top-down* method can be chosen, depending on the nature of the engineering problem at hand. Selection of appropriate methodology should also take into account:

(i) performance-based criteria related to the functioning of the system,
(ii) affordability of fabrication during the demonstration phase,
(iii) cost-to-performance criteria for efficient manufacture and sustainable resources during a product's manufacturing cycle.

This is not a trivial task for scientists and product developers, particularly since all three of these groups of criteria need to be satisfied simultaneously.

Another aspect of the philosophy of micro- and nanofabrication is the concept of integration levels. The terms hybrid and monolithic integration are used, with fully monolithically manufactured systems being the most integrated. From an industrial point of view, however, the latter is not necessarily the most favourable. It is the requirements of a specific product in a market place that determine the optimal approach.

One of the key innovations during the last five decades has been the controlled deposition of thin films. Solvent-based liquid, vapour or gas phase deposition processes are now readily available in the microelectronics industry. Thin-film technology is also an important driver for the definition of materials at the nanoscale. Tuning a material's properties allows the creation of novel applications. An introduction to nanofabrication is presented in Chapter 4 of this book. Presenting emerging nanofabrication techniques in the context of this book will allow us to draw additional attention to the paradigm shift that is now occurring in the world of production processes. Chapter 4 draws specific attention to the challenges of the positional assembly of nanoscale building blocks.

Unfortunately, at this moment it is difficult to give a clear definition of either micro- or nanotechnology. Thin-film technologies, for example, are already difficult to put into one category or the other, so it is important to clarify what one means when using the terms micro- or nanofabrication. In this context we distinguish nanofabrication from microfabrication techniques if the control of lateral dimensions can be realized precisely enough to result in features at the sub-micrometer scale. Within this framework, the structure height-to-line width ratio (aspect ratio) becomes a special measure of process performance in microsystems' product development.

As far as MST is concerned, it is easier to give a definition of the field of work instead of defining what actually micro- or nanotechnology is. MST leads to systems-on-chip devices, including their peripheral systems, containing electronic, sensing and actuating functionality packaged within a volume of a few cubic centimeters. Hence, specifically in the USA, this field is historically referred to by the name microelectromechanical systems (MEMS) technology. Sometimes the integrated electronic circuits, sensors and actuators are referred to as the brains, eyes and arms of artificial systems, an expression that has been used by Huff in his tutorial report on MEMS technology [1]. Optical switching and attenuation components are another large group of devices which benefit from the research and development into micro- and nanofabrication principles. The manufacture of microsystems involves a variety of precision engineering techniques, combined with silicon surface micromachining techniques. The first group of techniques (e.g., dicing, laser cutting, lamination, thermal compression bonding) enables the separation or assembly of components, while micromachining is dedicated to the integration of the three sub-functions from the electrical, mechanical and optical domain at the microscale.

With respect to the choice of materials, silicon is the most abundantly used material in miniaturization, although fabrication techniques should be evaluated for substrate materials other than silicon. A systematic overview of basic microfabrication techniques is given in Chapter 2.

The next three sections will give an overview of the content of this book. I also wish to present some guidelines on how this book may be used educationally as part of a course concerned with micro- and nanofabrication technology and its impact on industrial applications.

1.2 THE INDUSTRY–SCIENCE DUALISM

It is interesting that today, in the year 2011, we talk about knowledge societies. It was not long ago that productivity and industrialization were considered to be the key to a better health and education system, and hence a better life. In this section, we would like to discuss briefly the process behind the accumulation of knowledge, which is important for feeding existing industries such as raw materials industries (e.g., mining and farming), manufacturing, or services industries (dealing with law, medicine or the distribution of goods). Recently a new type of industry has been added to the three existing sectors: the knowledge industry, which deals with research, design and development. Knowledge creation is a skill one can teach. Children, teenagers and young adults all around the world are taught that knowledge is a driving force for building their careers. Adults are the workforce for the many successful businesses, including multinational corporations and governments. Industrial organizations cannot exist without mining accumulated knowledge or the application of business intelligence.

In contrast with an academic (pre-competitive) knowledge creation, however, industry points towards the economic valuation of knowledge: knowledge had to become a marketable product. Hence, it became a common practice to apply for patents.

A patent grants the owner the exclusive right to commercialize products or services. The idea or process being protected by a patent must contain an innovative step, and the disclosed information must be industrially applicable when the patent application is filed. Of course, if such an idea is disclosed it is already branded as intellectual property (IP), but a patent or trademark must be first registered and granted before the knowledge can be effectively traded on the market. This is a concept inherently different from science as an academic occupation. The right that one receives to commercialize the disclosed knowledge is a non-tangible asset of a business, which can be very valuable.

With respect to the development of micro- and nanofabrication techniques for industrial applications, it is important that we understand how IP is first generated and secondly traded. The latter is important if one wants to sell the accumulated knowledge through technology transfer. Not-for-profit organizations such as universities can be important trading partners in this knowledge industry. This approach may be easier understood if one realizes that a marketable, tangible product does not normally consist of only one invention. Several, often at different developmental stages, have to be combined to be able to form one product. Obviously, higher complexity will increase the number of components required, which means that the involvement of other market players becomes increasingly important for the effective realization of the product. For this type of development, a relatively new term has been introduced to business models: the concept of open innovation.

Although many companies will still be able to draw good profits from closed innovation, the complexity of systems in the modern world asks for more flexibility.

If you are not familiar with IP rights and their generation, the website of the World Intellectual Property Organization (WIPO) is a good starting point for finding out, and also for continuous updates on the procedures [2]. For details of the patent application process, and the legal rights attached to a patent, other sources are better, e.g., for Europe it may be helpful to check specific guidelines concerned with European patents [3, 4]. There are also guidelines on this topic for other countries and regions. Remember that ordinary people, as well as large organizations can apply for a patent, and many inventors use this instrument to pass on their knowledge to avoid the need for large investments. Large investments of the order of millions of US Dollars are very common for the initial steps of product development and scientific publication of the finding may compromise the inventor's commercial rights to it, but a patent can be applied for and maintained through various stages of this development process without significant costs. Also, inventors in small and medium-sized enterprises as well as researchers in the academic community can create businesses in this knowledge industry, by making use of patent protection as a business instrument by first attaining the commercial rights to their IP and then selling it to create their own knowledge-generating businesses. If financially sustained with the appropriate authorities, a patent is granted for 20 years, which in today's fast life-style is probably a reasonable head start over possible competitors.

How is IP initially generated? Of course, one immediately wishes to say: IP is generated by good ideas! However, that is not sufficient according to the description above, which suggests a patent is only granted if there is an inventive and an innovative step as well as an industrial application, which means, that quickly the inventor finds him- or herself in the area of technology. In the field of technology, the design plays a major role. Replacing the word *technique* in the field of technology with *design*, we can also formulate a guideline for exercising the knowledge presented in this book by a free translation of a Dutch quote taken from van den Kroonenberg lecture notes in 1998:

> *"Design is the process, which is directed to combine the knowledge of the existing laws of nature in such a way that it increases the efficacy of collection of knowledge in relation to another matter on the reason of need, demands and wishes of a human being." [6]*

How is a human being, an inventor, actually collecting natural laws to realize the process of knowledge generation, and how is an inventor making use of it in a design process? The guideline, unfortunately, does not give direct answers to these questions. This book is dedicated to the complexity of such design processes in the use of microfabrication techniques in industry.

1.3 INDUSTRIAL APPLICATIONS

Since knowledge from technology may have multiple applications across several different working fields the term "industrial applications" is specifically chosen to underline the potential of microfabrication for use in other research fields and marketable solutions. From a business point of view these multiple applications would be furthered by licensing agreements, which would allow the technology to be used commercially for a specific purpose by different industrial players.

To be able to clarify the definition of industrial players one needs to have a closer look at the creation of markets. Based on a simple definition in marketing we can call the total population "a market", but the "potential market" is the portion of the total population who have an interest in acquiring the product or service under consideration. In this definition the available market covers the sub-population that has sufficient money to afford the product. The latter may shrink to a qualified market, which represents the sub-sub-population that is eligible to buy the product. This qualified available market may subsequently be reduced to the target market, which is the market share that the company has decided to serve with its product (sub-sub-sub population). Finally, the penetrated market remains, which is the sub-sub-sub-sub-population of those in the market who have actually purchased the product. This final sub-sub-sub-sub-population is difficult to capture at the very early stages of a product development cycle, particularly for a novel and complex technology, which has yet to prove its worth. This is often referred to as the technology gap between science and industry. One way to approach this is the reduction of risks by developing a product over several stages that deliver a proof-of-concept for each stage. Early stage investments into a rising technology star, obviously, may promise very attractive profits.

Figure 1.2 conceptually presents the three possible market drivers, which divide the market population into three sub-populations (markets) that are

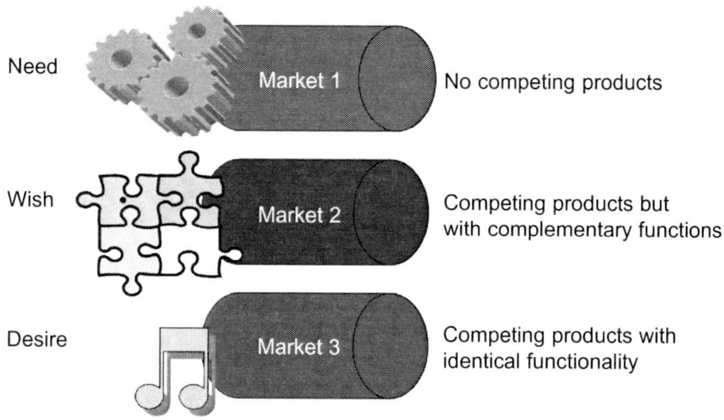

FIGURE 1.2

Drivers for gaining market shares.

driven by their specific needs, wishes and desires. At a very early stage of business it is difficult to propose that these markets could mix and merge. Currently, few publications exist that enlighten us on the mechanism of successful business development in a knowledge society [5]. I would like to suggest that at this initial stage of considering a technology–product match it may be possible to analyze these three drivers individually. With high-tech products, the three areas may merge eventually if a real need has been identified, and making products to satisfy this need is more likely to produce a satisfying return-of-investment than if a specific need is not identified.

1.4 PURPOSE AND ORGANIZATION OF THIS BOOK

This book is a study book. A study book should not only make you buy the book based on an interesting title but should also satisfy your needs; in this case, a need to learn more about microfabrication for industrial applications. This book is not a thesis on the definition of the economical value of microfabrication nor a dedicated technology handbook for specialists. The focus is micro-nanofabrication in innovative industrial areas, such as the field of (bio)medical applications.

Chapter 2 presents an overview of the basic process technologies of contemporary microfabrication. Chapter 3 extends this to broaden the range of materials being used, and considers processes that specifically aim at three-dimensional microstructures. Chapter 4 is an introduction to the terminology of nanotechnology, and an overview of the development of nanolithographic techniques that can pattern features less than 100 nm in width. Those which can produce large-scale, highly ordered pattern arrangements are of especial interest. The latter are often referred to as meta-materials.

Chapter 5 briefly reflects on the components and techniques of micromechanical transducers. Chapters 6 and 7 give an introduction to the application of microfabrication in the field of bio-chemical sensors and microfluidic Lab-on-a-Chip technologies, respectively. Chapter 8 provides the reader with a case study of microfabrication research dedicated to the drug delivery market. This example of desk research guides the reader through a commonly applied approach in the engineering sciences: knowledge accumulation and reduction of risk by applying a systematic design analysis. Finally, Chapter 9 gives some reflective comments and conclusions concerning the safety of micro- and nanotechnology and nanoparticles in particular. This last chapter also provides the reader with an overall conclusion to the book.

We will discuss fields of applications and introduce specific cases, which may have already entered the market, in the various chapters. Many examples of science-based innovation utilizing microfabrication can be found in the health sectors and a specific example is discussed in Chapter 8. Product development cycles are particularly long (>10 years) in the medical devices industry due to the necessity for devices that function robustly, to a high safety standard, because they are to be used in close vicinity to the body or inside it. Due to my personal research interest in the medical devices' sector, particular attention will be paid in the book to the opportunities for micro- and nanofabrication to play a part in the manufacture of novel implants, diagnostic and therapeutic systems. Nevertheless, some microfabrication applications will be also presented across other markets to indicate the variety of industrial perspectives which are possible.

The book contains case studies for students in the engineering sciences and for non-engineering specialists interested in this specific field of engineering. They reflect the current state-of-the-art of microfabrication in industrial applications, and seek to interest the student in the more advanced literature that is beyond the scope of this book. The book also provides an introduction for technology managers who are possibly experienced in business but less so with the terminology in the emerging field of micro- and nanofabricated industrial applications. For a specialist, e.g., a starting PhD student from the natural sciences or another monodisciplinary background, including the social or business sciences, who wishes to cross boundaries and enter the field of MST, this book may serve as an initial reference source on applications with a highly cross-disciplinary nature. Advanced microfabrication books, and the scientific and patent literature often discuss sophisticated theory of microfabrication technologies, making them less accessible for non-specialists. This book hopes to provide a starting point in an educational engineering course positioned at the end of the first year or beginning of the second year of a bachelor study program. Therefore, this book serves as an introductory source in the field of microfabrication technologies and its industrial applications.

References

[1] M. Huff, MEMS fabrication, *Sens. Rev.* **22** (2002) 18–33.

[2] http://www.wipo.int/portal/index.html.en, World Intellectual Property Organization, 2011 (accessed 04/04/11).

[3] J. Wild (Ed.), Patents in Europe 2008: Helping business compete in the global economy, *iam-magazine*, Globe White Page Ltd (2008).

[4] *How to Get a European Patent: Guide For Applicants*. European Patent Office, The Hague, The Netherlands, 2007.

[5] A.M. von Raesfeld Meijer, Technology transfer: Preaching to the converted or seducing the disbelievers, in M.K. de Laet (Ed.) *Research in Science and Technology Studies: Knowledge Technology Transfer, vol. 13, Knowledge and Society*, Amsterdam, ISSN 0278-1557, 2002, pp. 127–151.

[6] van den Kroonenberg, Methodisch Ontwerpen, Technische Hogeschool Twente, Afdeling der Werktuigbouwkunde, Vakgroep Ontwerp- en Constructieleer., 1998, page 1, identification number WB.83/OC-12461.

CHAPTER

Basic Technologies for Microsystems

2

CHAPTER CONTENTS

This chapter aims to make the reader comfortable with the tremendous amount of fabrication techniques that have found their way into applications since the onset of microtechnology in the 1950s and 1960s. Although there

Microfabrication for Industrial Applications. DOI: 10.1016/B978-0-8155-1582-1.00002-2

exist many excellent texts and reviews on the art of microfabrication, this chapter also introduces the reader to the nomenclature of micro- and nano-fabrication techniques as used in the remainder of the book and in science and industry.

Although many of the processes in use were initially derived from microelectronics, design engineers quickly realized that this approach restricts the possibilities for mechanical, optical, chemical and biological functionality. Hence, this chapter starts with an explanation of the primary techniques that were first performed with silicon and then moves on to the technologies which allow the use of other materials. The introduction of new materials has demanded the development of new microfabrication techniques, which we like to refer to as advanced microfabrication technologies. These are described in the next chapter. However, an introduction to non-silicon-specific techniques is given here, too. Furthermore, processing and application examples of the basic techniques are illustrated in a set of selected references.

2.1 PHOTOLITHOGRAPHY

Lithography generally involves the transfer of a pattern from a master into a pattern transfer layer, the resist, which is then used for subsequent pattern transfer onto a working surface, e.g., silicon. There are several different lithographic pattern techniques, of which photolithography is the most common.

This technique uses a resist layer that is sensitive to ultra-violet light. Accordingly, the light source is an ultra-violet (UV) source, which illuminates the resist through a specific master. The light spectrum of a mercury lamp has a UV peak at 365 nm. Combining a mercury light source with the necessary tooling for holding and moving the master (the mask) in a precisely defined way against the working surface allows the patterns to be aligned very closely. This combination of optics, precision engineering, and sensory control and feedback has resulted in one of the most essential production apparatuses of the microelectronic industry. They are known as mask aligners and are provided by specialist manufacturers. ASML, which is based in the Netherlands, is one such company. Any microfabrication process starts with a work piece, called the substrate or wafer, which may only serve as a handling platform, or may become a part of the integral device design. Wafers can be made from different materials, of which silicon is the most established.

During lithography, the resist, a polymer-based, light-sensitive compound of defined viscosity, is deposited onto the wafer. This may be performed simply by dispensing a puddle of solvent-dissolved resist onto the surface of the wafer and then rotating the wafer at high speed (spin-coating). The initial acceleration, reached speed, spin time and viscosity of the resist will define the thickness and uniformity of the coating on the wafer after the solvent has evaporated. Upon exposure to the light source, the molecules in the resist either cross-link (negative tone resist) or undergo scission (positive tone resist). The latter is often preferred because of the direct relationship between exposure intensity and rate of dissolution in the subsequent development step. These two processes of exposure and development are very similar to the same processes in photography. Negative tone resist requires an additional catalytical reaction, which is often initiated and accelerated by a moderate increase in temperature, making this resist system more complex. However, both types are widely used, with selection depending on the specific purpose the resist layer has to serve in the processing sequence.

There are a number of resist materials available commercially. The term 'resist' is used because this layer is used as a masking layer for the next etching or deposition. Hence, the resist allows the removal or addition of material to or from a specific region of the wafer. In the case of an additive process, the resist will be dissolved and carry the deposited layer from these regions with it. This process is called a lift-off process. The resulting boundaries of the patterns in a lift-off process are not as sharp as in etching processes and it has therefore its limitations. It is often used when materials that are known to be difficult to etch, e.g., ceramic thin films or metal films such as tantalum or platinum, have to be patterned. Hence, the resist may be patterned and used as a sacrificial material during selective deposition processes. For more details on these working principles the reader is referred to good text books in the field of microtechnology, for example, *Fundamentals of Microfabrication* edited by Madou [1].

Besides a layer of light-sensitive resist, photolithography also requires a special master, a mask, which blocks (absorbs) the light in defined regions and is transparent in the remaining regions. These masks are fabricated by a direct-write method using either electron or laser beam pattern generators.

The pattern is then transferred from a mask to many wafers by a print-like process, using the mask aligner apparatus mentioned above. An example of a mask aligner, operated in a dedicated laboratory (cleanroom) is depicted in Figure 2.1.

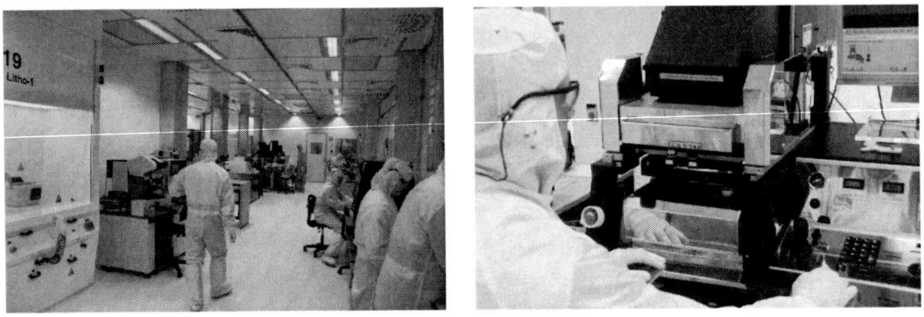

FIGURE 2.1

NanoLab NL cleanroom facility at MESA+ Institute for Nanotechnology (left) with mask aligner apparatus for UV lithography in operation (right). (Images courtesy: S. Schlautmann, University of Twente, 2011.)

The reader is referred to Madou [1], and Cui [2], for further details of lithographic techniques. Direct-writing of a pattern onto a resist transfers a computer-aided design (CAD) into the resist on the mask blank, and subsequently the pattern is transferred by etch-through into the light-blocking material. The mask blank, therefore, must be UV transparent and in most cases a borosilicate plate with a chrome layer as an absorber is used [3, 4]. The mask-making process is also a lithographic process, however, due to its serial write strategy it is slow. This makes it too expensive for use in large-volume production.

The selection of masking procedures and the choice of materials and techniques will influence the accuracy of the final device's geometry, and, therefore, the device performance. Low-cost masks also exist. These are processed by printing the CAD design onto an ink-jet transparency, which is subsequently transferred to a standard mask blank using a mask aligner or UV-light box. However, resolution and line width roughness are restricted by the resolution of the printer.

The CAD files are often generated in specially designed software tools that provide functions for drawing a mask layout, which can be converted by the pattern generator into a latent image in the resist on the mask blank.

Lithographic and more recent plasma etch methods are commonly used to remove material, leaving strips, lines, dots and other shapes on the substrate with well-defined profiles and dimensions in the individual device components. The resolution achievable in photolithography is restricted by the diffraction of light and therefore it is limited in its lateral dimensions. The degree of resolution of a lithographic process refers to the narrowest distance that can exist between two lines such that they can still be transferred as two separated lines (fully resolved lines).

Other fast and novel direct-write strategies for wafer production have been developed recently. Novel parallel electron or ion beam techniques allow features to be copied at increasing manufacturing speed by multiplication of the active write element. Unfortunately, none of these techniques are yet on the market and are so far only available in research [5]. Such methods will overcome the limitations of optical diffraction, so extending the application of top-down patterning processes far into the sub-micron domain.

An overview of novel nanolithographic techniques is given in Chapter 4. Laser stereolithography or non-lithographic direct shaping methods (micro-milling, laser ablation, 3D-printing) may be used as direct-write fabrication techniques for less demanding dimensions. These techniques are also very popular for the fabrication of microfeatures during prototyping in research and development, and for small volume or single piece production. We will discuss some examples in Chapter 3.

2.2 THIN FILMS

Silica and silicon nitride are classical thin film materials, functioning as dielectrics and capping materials in microelectronics. Metals such as copper, gold, platinum, titanium and chromium are another type of materials used to make thin films forming conducting, masking, sacrificial or adhesion layers. A comprehensive overview of processes for the deposition and selective etching of thin films can be found in many texts, e.g., Vossen [6].

The range of materials used in microfabrication is extensive. Besides these conventional materials, various polymers, oxides and III-V-based semiconductors as well as many hybrid materials have been investigated for the production of integrated systems [7–12]. Many new, thin film systems have been designed, based on novel instruments, which are capable of the controlled manipulation of nanoparticles or even atoms during their deposition. These materials have very advanced properties, but they can still be deposited by fairly simple techniques from solution. A range of standard deposition techniques for thin films will be discussed in the following two sections.

2.2.1 Wet Deposition Techniques

Deposition of a paint occurs by evaporation of the solvent from the suspension. In traditional wall coatings (i.e., paint) the thickness of the final coating and its uniformity are obviously not very critical parameters. However, in microfabrication they are. Despite this major difference, the principles of deposition which are used for dipping, spin- or spray-coating in patterning at the micro- and nanoscale remain the same. The film-forming compounds are dissolved in a solvent. When this is spread onto the handling platform (substrate, wafer) the solvent starts to evaporate and the film is formed. Photoresist, e.g., Microposit® S1813, may be deposited by spin-coating [1, 13, 14]. This process can be supported by a baking process usually called a soft-bake at a modest temperature around 90°C.

As mentioned previously in Section 2.1, a resist system may act either as a positive or a negative photoresist. In a positive resist, the area modified during the lithography step is removed from the substrate. This creates openings that allow to act on subsequent process steps. In negative resists, the lithographically-manipulated region remains on the substrate.

Besides masking, novel nanotechnology-based coating processes also allow the formation of more complex thin films by relatively simple deposition techniques which exploit such processes as polymer-demixing of multiphase systems (two or more phases are dissolved and remain in the matrix of the material) and may include nanoparticles (e.g., nanosphere lithography). When spreading the hybrid organic–inorganic nanocomposite suspension onto a substrate, e.g., by spin-coating, the building blocks in the material systems reorganize to minimize the energy state of the film. This can also explain why room temperature annealing over a period of a few hours enhances the ordering of such matrix systems after deposition of the film. Single layers of different properties may alternate during deposition, and form a so called layer-by-layer film by self-assembly, which may not only have the alternating original functions of each individual layer, but also completely new properties [15, 16]. An introduction to such novel techniques and special thin-film materials can be found in Chapter 4, Section 4.2.

2.2.2 Vapor Deposition Techniques

The reader is referred to references [17], [18] and [19] for further details on this topic, since this discussion is intended as an introduction.

Thermal evaporation of metals having low to moderate deposition rates (10–100 nm/min) is regarded as one of the traditional physical vapor deposition (PVD) techniques. It relies on the sublimation of material onto a substrate in a vacuum. In industrial applications, the so called sputtering process is often preferred to thermal evaporation, because of better step coverage, a wider choice of possible materials (includes dielectrics) and often higher deposition rates (up to a few 100 nm/min), while forming dense films with an appropriate adhesion quality. Sputtering forms ions from an inert gas, which are accelerated against a target material. This then goes into the plasma phase and deposits onto the substrate (the work piece) as a film. The process was developed to meet the requirements of large scale integration (LSI) in the 1970s and 1980s. Film properties are actively controlled by manipulating parameters like bias, pressure and substrate heat. These control mechanisms allow for higher reproducibility, a better device optimization and result in adapted material properties. Depending on the properties of the film-forming compounds, it is also possible to incorporate special properties, e.g., UV sensitivity into the film, specifically investigated for advanced lithography resist materials [20].

Chemical synthesis, using precursors in the gas phase at elevated temperatures is also possible. These are called chemical vapor deposition (CVD) techniques. The elevated temperatures that are necessary restrict its use to wafer materials that can withstand these deposition temperatures, and match to a certain extent the thermal expansion coefficient of the deposited film. If these processes are conducted at subatmospheric pressures they are called low pressure chemical vapor deposition (LPCVD). Nearly all CVD processes are carried out at low pressure, to avoid unwanted gas-phase reactions and improve film unformity at elevated reaction temperatures. If a plasma is used to provide energy during the material deposition process, it is called plasma-enhanced chemical vapor deposition (PECVD). These techniques are of interest in applications where a low substrate temperature and a high-density film is received. CVD techniques involving a plasma reactor are fully established in industrial production.

A variety of methods in which the plasma not only provides the energy source but also enables different modes of deposition are also available [21]. Parameters such as the total reactor pressure, frequency of the rf-excitation, rf-power and growth temperature all influence the film morphology. Although other film-forming processes exist, physical and chemical vapor deposition

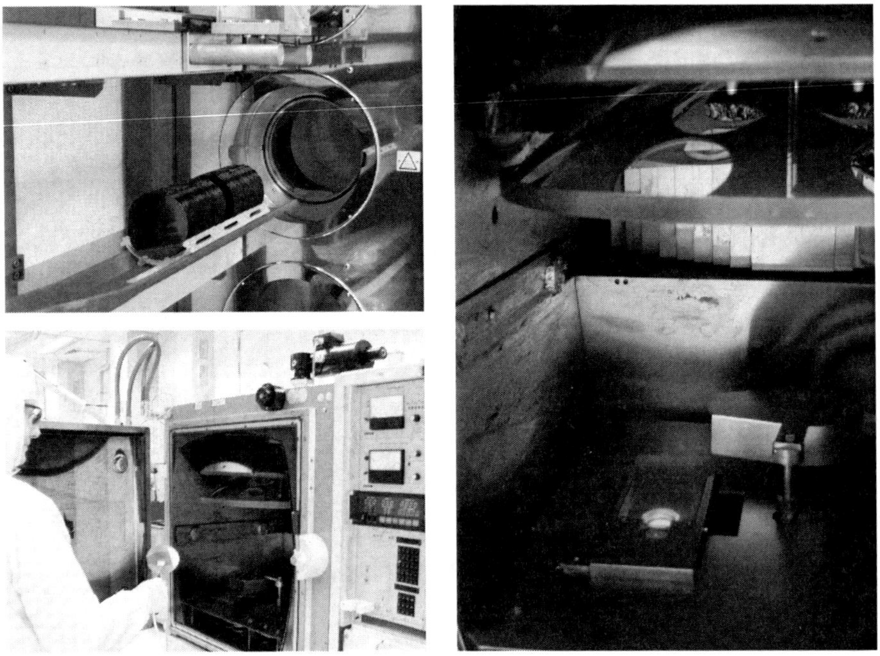

FIGURE 2.2

NanoLab NL cleanroom facility at MESA+ Institute for Nanotechnology. Top left: low pressure chemical vapor deposition (LPCVD) oven. Bottom left: loading a thermal evaporator. Right: showing the interior of the vacuum chamber of the evaporator with the wafer carousel on the top part of the figure and the crucible at the chamber bottom. (Images courtesy: S. Schlautmann, University of Twente, 2011.)

techniques are the 'workhorses' of industrial thin film applications. Figure 2.2 shows two such thin film generating apparatuses.

Pulsed laser deposition (PLD) is a recent technological development, and an example of the shift in research to nanotechnology [22]. A new material can be composited based on an understanding of reaction kinetics in the gas phase.

This technique can be used to design thin films with unique properties, e.g., magnetic, superconductive, negative thermal and electrical resistance materials and multilayers. Thin films clearly possess the potential for innovative micro- and nanofabrication applications, and for deepening our understanding of their physicochemical properties at the nanoscale level. These can be investigated by scanning tunneling microscopy and other novel imaging techniques that resolve atomic interaction phenomena stirred by the deposition process. New thin film technologies also involve a search for novel materials. It is therefore very exciting for researchers and developers to discuss thin films in the broader context of the innovation stimulated by nanotechnology.

2.3 SILICON MICROMACHINING

Silicon is one of the fundamental materials of the semiconductor industry and was used nearly exclusively in the early applications of microsystems technology (MST). Even now silicon is the preferred material for the initial testing and integration of devices because its patterning techniques are very well known. Silicon offers a high degree of monolithic integration capacity and has very good mechanical properties. It remains a very attractive material because of the high quality of wafer finishing achievable. This also makes it an interesting material for mechanical sensor applications, since they often rely on precise bonding at the wafer level [23–25].

The fact that silicon wafers are a relatively inexpensive commodity allows researchers in commercially yet less attractive fields access to this base material and to run a variety of well-defined processes. However, this can bias the choice of fabrication techniques, and researchers often look for a work-around in their device designs to meet silicon micromachining techniques. This approach sometimes precludes a successful (cost-efficient) implementation of a new technology. On the other hand, silicon is indispensible for making demonstrators of a novel functional mechanism at small scales although the material may be substituted for alternatives if the device goes into full-scale production. The controlled deposition of thin films and their partial removal by wet or dry etching play an important role in silicon micromachining.

Wet chemical etching is a fairly simple technique to perform, and the associated equipment costs are relatively low [26]. However, established silicon micromachining techniques are restricted to etching in layers up to a few micrometers in thickness, which are inadequate for some device criteria. The increasing demand for three-dimensional device structures has motivated many research groups to develop specific high aspect-ratio technologies with high precision [27–30]. Such techniques may include deep dry etching and three-dimensional masking in combination with anisotropic wet etching in silicon [31]. In reality, a microfabrication sequence can involve a great many processing steps to produce just one functional component of a system, e.g., microneedles or comb-like shaped actuators (comb-drives). Further details of the fabrication sequence for a range of devices and systems will be described in Chapter 5 (micromechanical transducers), Chapter 6 ((bio)chemical sensors), Chapter 7 (Lab-on-a-Chip systems) and Chapter 8 (microdevices for medical applications). The following sections of this chapter will review the variety of basic techniques for microfabrication specifically in silicon. The information is based on Petersen's comprehensive introduction to silicon as a micromechanical material (1982) [25].

2.3.1 Etching

Wet Chemical Etching

Wet etching is a purely chemical approach for the removal of layers or bulk material, the following processes are recognized as standards.

Buffered oxide etch (BOE) contains $NH_4F/HF/H_2O$ and is used for the etching of thermal oxide, LPCVD oxide and PECVD oxide, or low temperature oxides (LTO). BOE patterning can be performed using, for example, a positive resist (e.g., AZ1500 series from AZ Electronics Materials plc) with a defined thickness of 1.2, 1.8 or 6.5 μm [32]. Tetramethylammonium hydroxide (TMAH) or potassium hydroxide (KOH) silicon etchant is used for the crystal-oriented anisotropic etching of mono silicon. Thermal oxide or LPCVD nitride masks are commonly used for patterning.

A so-called HF-dip is used for etching thin silicon oxide (gate oxides, native oxide layers). A phosphoric acid (H_3PO_4) etch can be performed for uniform stripping of silicon nitride layers (LPCVD nitride, PECVD nitrides).

Aluminum etch is a special mixture of acids, used for etching of aluminum layers (e.g., interconnection lines). Polysilicon etch is a mixture of HF and HNO_3 and is used for isotropic etching of polysilicon layers.

Many other reagents are used for special applications, including special etchants for metal layers (Au, Cr, Ni, Cu, Ti). Figure 2.3 depicts an example of a standard wet-etch bench (top left) including a special rinsing system, a quick-dump rinser (bottom left) and schematic representations of two types of etching mechanisms in silicon (right).

The chemistry of the reactions is isotropic in nature, unless selective etch rates are established due to crystallographic planes or change in materials

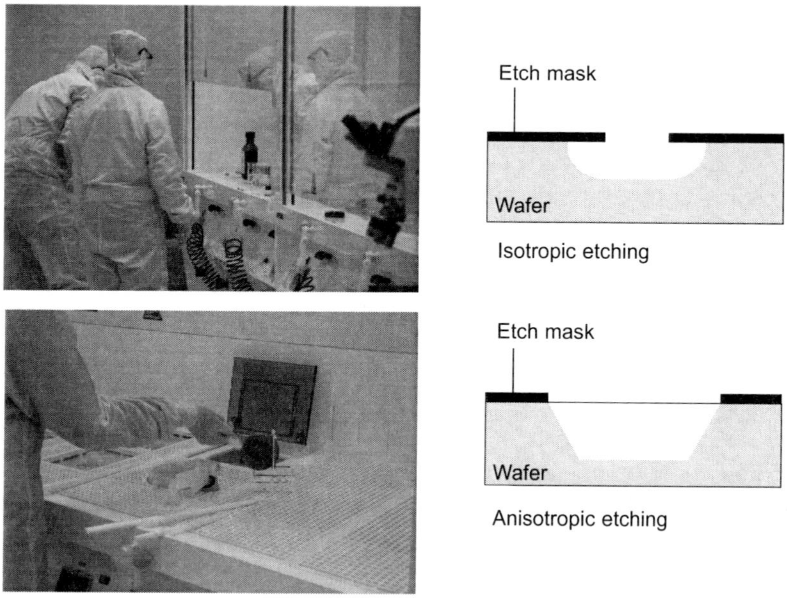

FIGURE 2.3

NanoLab NL cleanroom facility at MESA+ Institute for Nanotechnology. Chemical etch acid/solvent wet bench (top left) and wafer handling at the quick dump rinsing station (bottom left). (Image courtesy: S. Schlautmann, University of Twente, 2011.) Schematic wet etching profiles in silicon (right).

(e.g., using etch-stop layers). However, isotropic (wet) chemical etching frequently suffers from instability of the masking layer at the edge of a pattern, due to lateral under-etching. Thin, overhanging parts of the mask can break off, thereby increasing the line width roughness. Isotropic wet etching is therefore restricted in resolution, and produces rougher, thicker lines than dry processing. Anisotropic wet etching differences in etch rates of different crystallographic planes is far less sensitive to line width roughness, but it may still suffer from delamination of the masking layer. Current design tolerances, and also the hazards involved when working with and disposing of large amounts of hazardous liquid-based chemicals, make these processes less attractive for industrial production than dry etching in vacuum systems, despite their simplicity. A comprchensive overview of etch rates for micromachining processing is given by Williams and Muller [33].

Dry Etching

Many available dry-etch processes bombard the wafer with ions through a masking layer. Some of these techniques can etch extremely deep features with vertical side walls, which will be described in more detail in Chapter 3. Here, an introduction is given to this ever more important technique of microfabrication. Unlike directional wet etching, these processes do not rely on the orientation of the crystal structure (lattice) but the selectivity of the etch rates of the materials used as masking layer and work piece. Selectivity is particularly restricted if physical ion bombardment is the major mechanism of the etching process (e.g., ion milling).

Basically, these ion etching processes use a parallel plate configuration to make a plasma in an inert gas under reduced pressure. If specially shaped electrodes are used to confine the plasma, and reactive gas species are used, then the etch rate and selectivity can be enhanced and the etch profiles can be tuned. Figure 2.4 summarizes the events which occur in a conventional reactive ion etch (RIE) process, as described in a variety of microfabrication processing handbooks [34, 35].

The essential features of RIE are:

(a) low pressure,
(b) anisotropy (preferential etching in a direction normal to the substrate surface), and

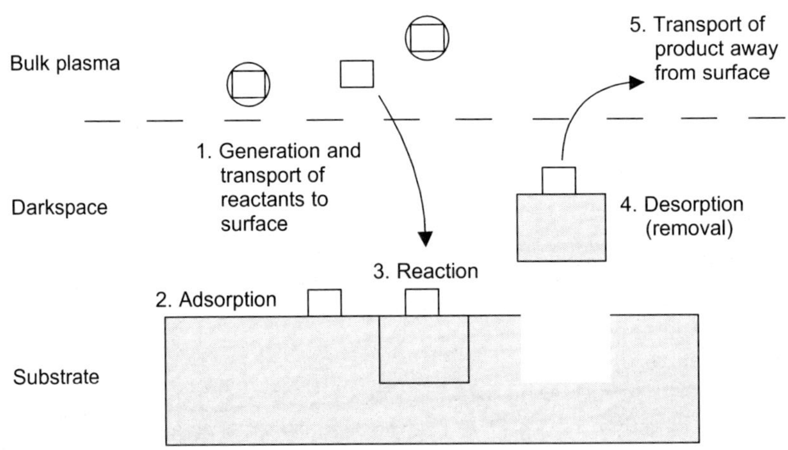

FIGURE 2.4

Essential events during reactive ion etching. (Redrawn from reference [34].)

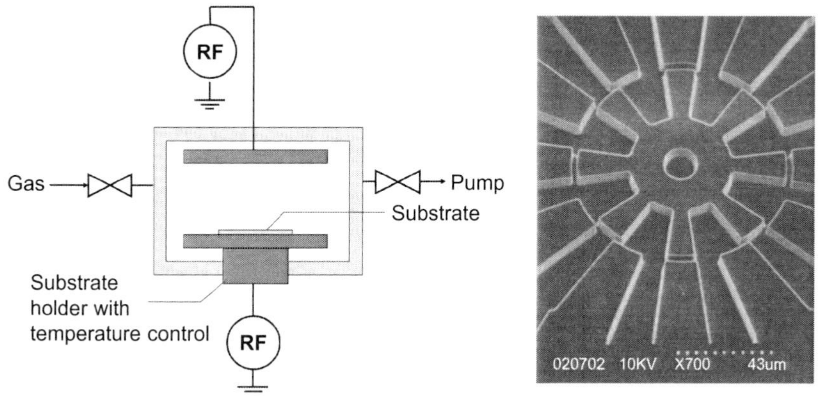

FIGURE 2.5

Schematic of a parallel plate plasma etching apparatus (left) and example of reactive ion etching result in silicon (right, reproduced from reference [36]. Copyright ©1995 The Electrochemical Society).

(c) selectivity (the etch rate ratio between the material to be removed and the masking material).

The RIE etch mechanism further includes:

(a) chemical etching for the creation of volatile products by reaction between the work piece material and the etch gas (e.g., fluorine-based gas) and
(b) sputtering, which refers to the interaction of ions with matter during the bombardment with chemically inert ions (e.g., Ar^+) in the plasma.

Figure 2.5 depicts the schematic of a parallel plate etcher configuration and one example of an RIE-etched microstructure in silicon.

2.3.2 Surface Micromachining

Various techniques have been developed to pattern different layers by surface micromachining. The principles of wet etching rely on the chemistry of dissolution of materials. Most of the processes are isotropic in nature and can be performed either in the liquid or the vapor phase. As mentioned in the previous section, dry processes are more complicated because they require a vacuum system and involve the generation of a plasma by electrical discharge, meaning that they need complex control equipment to modulate process parameters.

In dry etching processes, the use of a reactive gas species allows highly selective etching of the work piece in the presence of a masking layer. However, the specific gas chemistry, substrate temperature and field strength must be tuned to each material and etching application. In the case of plasma deposition processes, the gas mixture (i.e., composition) can be used to modify the layer properties of thin films. Patterning for many microelectromechanical devices has been demonstrated on thin films of polysilicon (PS). Micromotors with an electrostatic actuation principle were one of the early examples of successful polysilicon surface micromachining. This application of surface micromachining is introduced and illustrated in Figure 2.6. Figure 2.7 shows an etched and surface-released machined gear wheel that was fabricated using this processing sequence [37]. In general, fluorine-based plasma processing of silicon is selective against silica, which can be patterned either by RIE or in buffered HF. Other standard, dry etch, gas mixtures are chlorine- or bromine-based. Reference to a variety of process recipes can be found in various handbooks [34, 35].

Surface micromachining utilizing polysilicon (PS) is a key technology or so-called standard microfabrication process, now used to manufacture an extensive variety of devices. Micromotors were made possible by surface

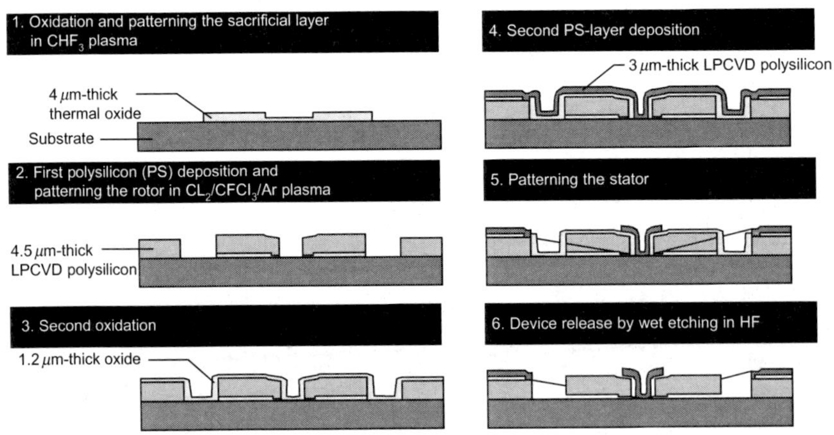

FIGURE 2.6

Integrated fabrication of a microgear-train realized from polysilicon. (Redrawn from reference [37].)

FIGURE 2.7

Cross-sectional view of a typical structure resulting from the process sequence depicted in Figure 2.6, a gear-train with the largest gear of a radius of 120 μm. (Reproduced from reference [37] with permission ©IEEE 1988.)

micromachining techniques that define sacrificial layers by means of which a structural feature can be released from the substrate. This technique is also commonly applied in the design of resonant microdevices, as discussed in Chapter 5. Sacrificial silica processing can be also applied to smooth and clean surfaces, and to generally reduce the effects of friction for mechanical moving parts and in devices for media transport (e.g., microfluidic chips). Silicon surface micromachining is of less importance for integrated optical waveguides in the visible range, but it plays a key role for the integration of micro-opto-electromechanical systems (MOEMS) and other photonics devices which operate at longer wavelengths. In optical devices, the surface or side wall roughness is directly related to loss of reflectivity in mirror devices. This makes the definition of line edge and surface roughness crucial for optical applications. Of course, in the ever-increasing demand for electronic integration reducing line width roughness is also important in electronic devices.

Surface micromachining is a delicate process with a relatively low device yield. However, process optimization produced a commercially attractive manufacturing process in the early 1990s. A number of silicon manufacturing foundries were established, and introduced a service for silicon processing and MEMS manufacture on demand, including multi-user MEMS processes (MUMPs®, where different users share device space on one wafer). This sharing of space is made possible by designing the devices according to a design routine which is supplied by the manufacturing foundry as a standard and thus at a reasonable price. Some of the players who first offered these foundry services were Sandia National Laboratories and Bosch. MEMSCAP, a French leader in the MEMS service industry, provides both standard and custom products to their clients. The MUMPs® program has run since 1992, following previous success at Cronos, a JDS Uniphase division. MEMSCAP acquired Cronos in 2002 and now it runs MUMPs® as a polymer, metal and silicon-on-insulator (SOI) process in addition to their standard complementary-metal-oxide-semiconductor (CMOS)–MEMS line.

2.3.3 Silicon Bulk Micromachining

A wide spectrum of machining techniques has emerged for the patterning of bulk silicon: from more conventional wet etching to complex anisotropic dry etching and novel, focused energetic beam processes. Once a new processing instrument is designed novel techniques can be derived for its use. Techniques are developed, and their specific patterning capabilities are researched. Further development produces a technology for the manufacture of components and systems.

Originally, many components were simply miniaturized versions of designs existing at a larger scale, hence, to begin with, silicon micromachining offered high precision but also dictated certain design criteria. A set of standard technological processes has evolved, around these design criteria as process modules or steps, which can be used interchangeably. The microfabrication expert will generally talk about a new process technology when a certain set of such modules are put in a new sequence to allow a novel geometrical structure to be manufactured – but the individual steps will have been known for sometime. An example of this is the development of technologies for the manufacture of microneedle arrays [31]. Considerable effort is currently devoted to developing industrially robust methods for manufacturing ever-smaller devices. As with processing at the macroscale, many process developments are related to the efficiency or yield of the process.

In current bulk microprocessing, the designs for a specific functional element may vary a lot from application to application, requiring specific tuning of the process parameters (substrate temperature, gas species, plasma energy, process pressure, rf-bias). Process selection and optimization is therefore based on the specific purpose of a device. This approach results from the fact that the same type of device can be made by different, patentable technologies resulting in unique properties during the device application.

As mentioned previously, patterning can be performed either by parallel or by serial processing. A mass-market approach using parallel, or batch fabrication is more desirable from a commercial perspective. For example, deep reactive ion etching has become very popular since the development of the "Bosch process" for silicon etching resulting in a high quality and fast etch rate [38]. These processes can now be bought-in from various suppliers (microsystems foundries), e.g., Silex (Sweden), LioniX (The Netherlands) or Tronics-MST (France). Foundries belong to the service industry and generally do not develop their own microdevice. Bosch etching and wet anisotropic chemical etching are the microfabrication techniques in most common use at present, and are introduced in the next two sections.

Anisotropic Wet Chemical Etching

Silicon has been exploited to an enormous extent as a micromechanical material. Simple shapes like V-grooves and cavities may be produced in silicon by selectively removing material from the substrate [1, 26, 39–42]. Although the concept of silicon etching has been described in the aforementioned sources, a short overview is presented here, concentrating on successful industrial applications, such as the etching of micromachined mass-spring systems for accelerometers, or the diaphragm-based pressure sensor.

To produce such systems an etch-resistant material (e.g., thermal oxide) is deposited or grown on the surface of the substrate, and patterned by photolithography and wet etching. As mentioned briefly above, there are many chemical etchants for silicon, but ethylenediamine perizine (EDP), or less toxic solutions like potassium hydroxide (KOH) and tetramethylammonium hydroxide (TMAH) are well established wet silicon bulk etchants.

Figure 2.8 depicts the type of vessel that is used for the process. The etchants in the double-walled vessels are temperature-controlled by a water bath recirculation system. Here, several 4″ wafers are vertically mounted in

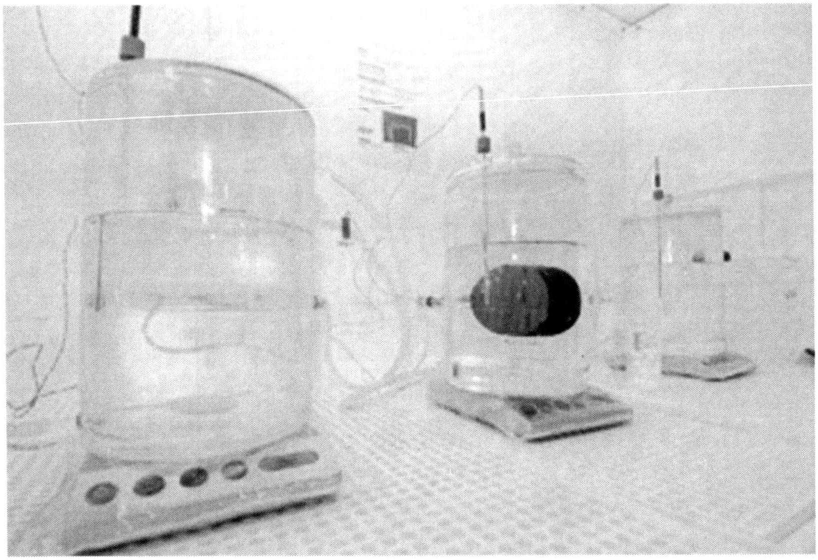

FIGURE 2.8

NanoLab NL cleanroom facility at MESA+ Institute for Nanotechnology. Examples of a typical vessel used in silicon bulk micromachining by KOH-etching. (Image courtesy: H.A.G.M. van Wolferen, University of Twente, 2011.)

a special quartz-glass rig. Their arrangement allows several wafers to fit into the vessel, and allows the hydrogen gas that is evolved during the etching process to escape from the structures. Etchant selection depends on the application. Etch rates are approximately 1μm/min, and etch times of 2–3 hours are fairly normal in these processes.

Timed etching, or etch stops such as a boron-doped silicon layer may be employed [43, 44]. The crystal-oriented etch characteristics of silicon in anisotropic etchants depends on the fact that the etch rate decreases dramatically when it reaches an inhibiting crystal plane [45–47]. Special simulation software has been developed to predict etch shapes. The high etch rates of so-called convex corners must be treated with particular care. An example of a silicon structure and its possible fabrication are shown in Figure 2.9 and Figure 2.10. More advanced information about the wet and dry etching of silicon is presented in an excellent review paper by Kovacs et al. [48].

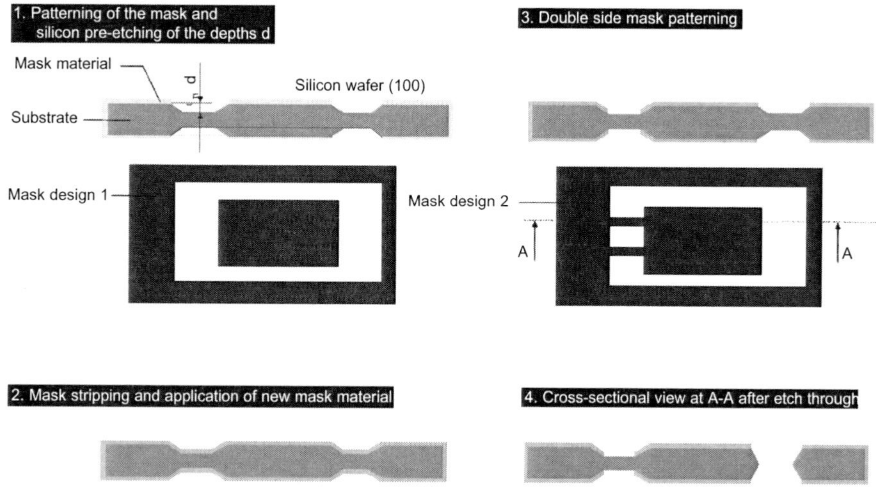

FIGURE 2.9

Process sequence for bulk etching with aligned double-side exposure.

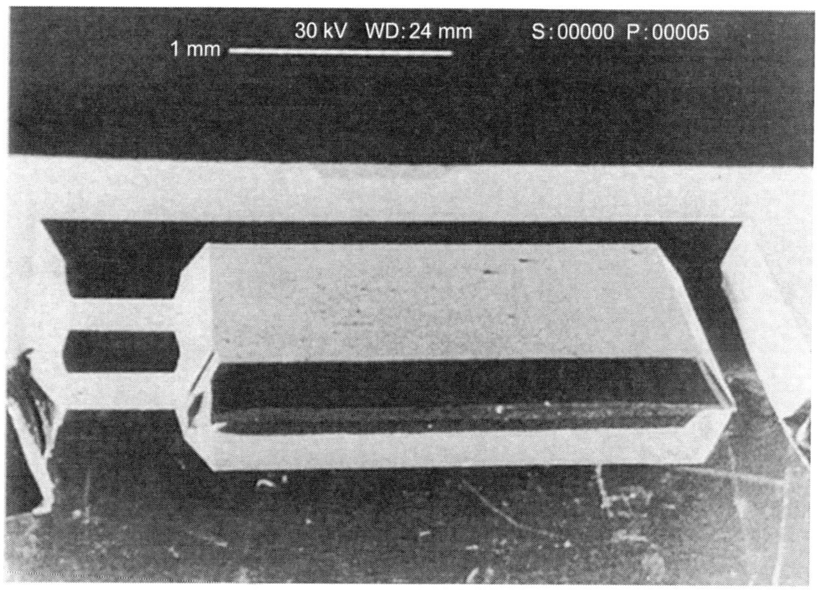

FIGURE 2.10

Seismic mass of an acceleration sensor realized by silicon anisotropic wet etching. (Reproduced with permission from reference [49]. Copyright ©2000 Springer Science+Business Media, LLC.)

Bosch Process

Reactive ion etching by the *Bosch* processing technique utilizes an inductively coupled plasma (ICP) source and alternating gas chemistry for the controlled steps of reactive ion etching (RIE) and deposition. Such an alternating process provides selectivity between silicon and the photoresist of 100:1, or over 200:1 if oxide is used as a masking material. These high ratios are obtained by using a plasma source which generates a very high plasma density but at the same time allows a high control over the ion bombardment energy. The use of fluorine-containing etch gases such as SF_6 (etching) and C_4F_8 (passivation), means that the *Bosch* process can coat the side walls, and also opens the bottom of the structure, by highly directional ion bombardment. Passivation and etching cycles are automatically switched every 5 to 15 s. Figure 2.11 shows microstructures etched by the Bosch process.

FIGURE 2.11

Typical example of a microstructure generated by the Bosch process. (Image courtesy: J. G. Bomer, University of Twente, 2004.)

2.4 INDUSTRIALLY ESTABLISHED NON-SILICON PROCESSING

Microsystems rely on shaping many materials into structural elements. During the last two to three decades many materials other than silicon have been used, either in surface micromachining, or as a thin film, or for components produced by bulk micromachining. Examples include piezoelectric single crystal quartz, photostructurable glass or plastics. A diverse range of materials has been investigated for use in this area. The following subsections illustrate basic fabrication principles that are often used in industrial applications.

2.4.1 Quartz Etching

Single crystal (SC) materials, such as single crystalline quartz, may be also patterned anisotropically in a similar way to silicon. Selective wet etching of specific crystal planes offers efficient removal rates using chemical etchants and can deliver high precision. SC-quartz wet etching has been widely investigated for the manufacture of frequency-control timebases as well as for a variety of sensors (force, strain, acceleration, temperature, pressure, to name a few). However, its application in fields other than tuning forks is rare. The complex crystal structure makes prediction of etch shapes more difficult than it is possible for silicon.

Quartz has been used as piezoelectrical material for micromechanical resonator devices. However, with respect to multifunctional capabilities quartz' major advantage is its optical transparency and its use has not yet been fully exploited for sensing and actuation through use of the piezoelectric effect. Also, quartz wafers can be cut along a symmetry axis, as such that a tuning fork device may be aligned in such a way that the symmetric fork design can compensate for temperature effects in the resonant behaviour [50, 51].

Bulk micromachining quartz for a micro-optoelectromechanical system is described by Sugiyama et al. [52], and shown in Figure 2.12. Miniature resonant tuning forks are used in the wristwatch industry. The steps of a double-side process are shown in Figures 2.13 and 2.14 [53].

2.4.2 Glass Wet Etching

Amorphous glasses, including fused silica, can be etched in a similar way to single-crystal quartz. Examples are the isotropic etching of borosilicate glass

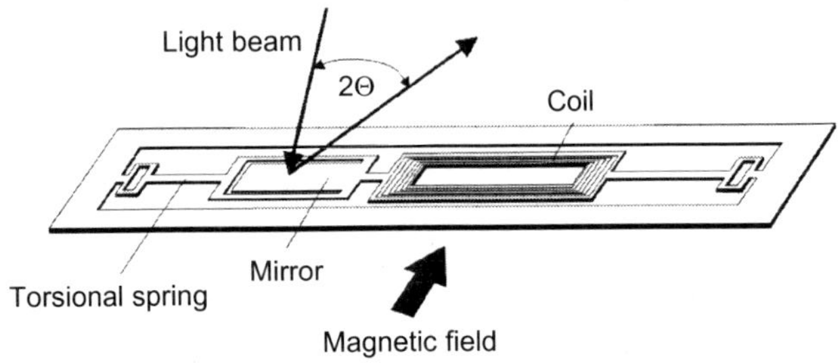

FIGURE 2.12

Quartz resonant scanner, electromagnetic scanning principle. (Redrawn from reference [52].)

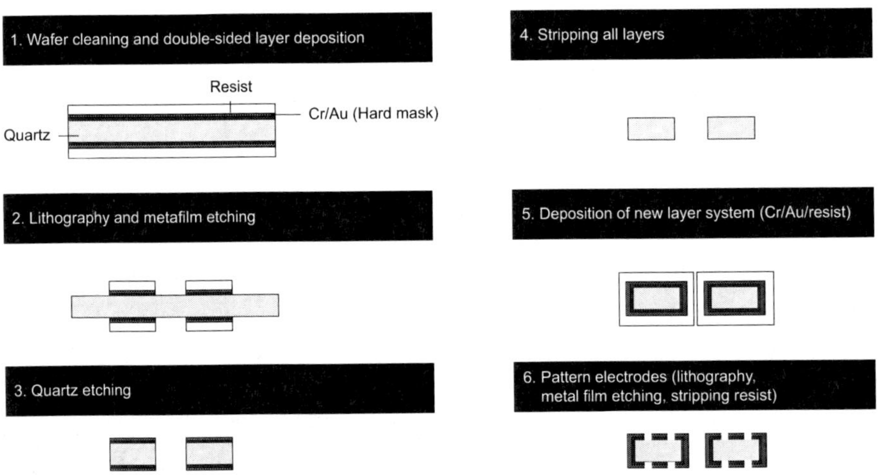

FIGURE 2.13

Process sequence for SC-quartz tuning forks.

in hydrofluoric acid (HF), in which a strip-opening in the masking layer will result in a hemispherical channel structure if sufficient etch time is allowed. Figure 2.15 demonstrates the application of this technique to the fabrication of microfluidic channels [54].

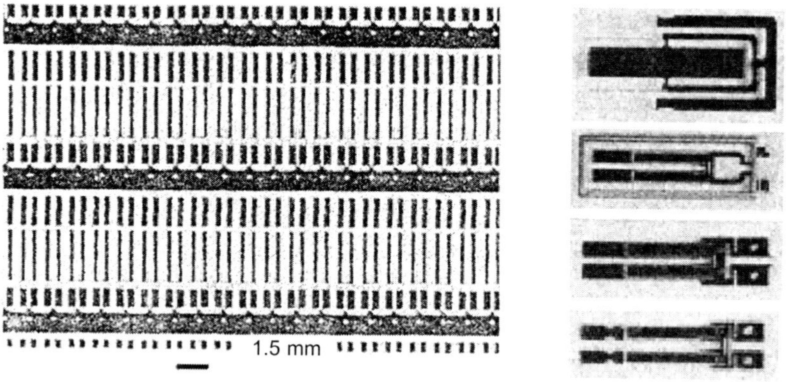

FIGURE 2.14

Left: quartz tuning fork resonators manufactured on a quartz wafer by photolithography. Right: various types of batch fabricated quartz resonators manufactured by the technique described above in Figure 2.13. (Reproduced from reference [53] with permission ©IEEE 1997.)

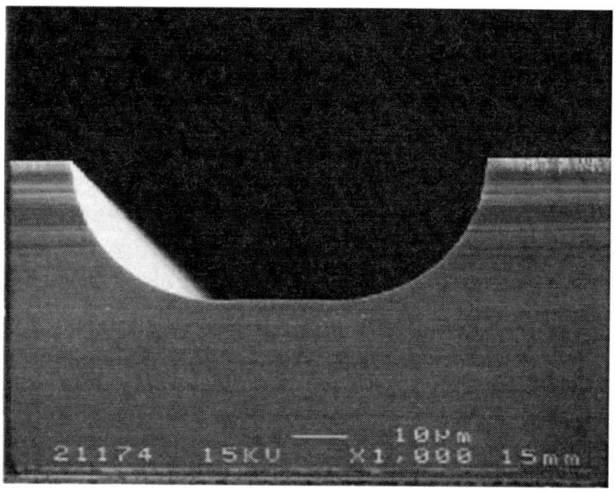

FIGURE 2.15

Scanning electron micrograph of cleaved edge of microfluidic channel etched in glass by HF. (Reproduced from reference [54]. Copyright ©1996 American Chemical Society.)

This process is very important for the fabrication of microfluidic-assisted biochips, which are a successful emerging market. Various chip manufacturers and high-tech suppliers, such as Philips and Agilent, have invested in Lab-on-a-Chip technology to develop their own applications in this field.

2.4.3 Photostructurable Glass

Amorphous glass is of great interest for the microfabrication of optical integrated microelectromechanical systems (MOEMS). It is possible to photochemically modify special glass by applying ultra-violet light [55]. This so-called photoetchable glass has been available for some time [1], and is now being used where structures with good thermal and electrical isolation are required. The resolution of the lateral dimension, however, is relatively low but high aspect ratios can be achieved. Pattern transfer characteristics are as follows: illumination of material up to 2 mm thickness; a convex feature with an aspect ratio of up to 40:1; the remaining surface roughness is in the order of 1 μm. Using Foturan® glass, various geometries can be realized, with the width of walls as low as 50 μm, the length of walls up to 100 mm and the diameter of holes (concave features) as low as 50 μm in substrates of 1 mm thickness. The material can be used in a similar way to a silicon or glass wafer, and has a surface finish quality appropriate for creating a bond at wafer level. Metal coatings can be applied for a variety of applications, too. One example of the application of photostructured glass is in the field of microreactors [56]. To pattern through 1 mm thick Foturan® plates a photo-energy density of 2 Joule/cm^2 is required. Silver clusters are formed within the exposed areas. The second step requires a heat treatment between 500°C and 600°C. The thermal energy causes crystallization of the glass around the silver atoms.

These crystallized regions can be etched up to 20 times faster than conventional glass (e.g., BK 7) and unexposed Foturan® in a 10% hydrofluoric acid solution at room temperature. Foturan® is a trademark speciality glass manufactured by Schott AG, Mainz, Germany. PEG3 is another photosensitive glass synthesized by the Hoya Corporation, Tokyo, Japan, and although these types of glasses are important their commercial activity has been discontinued for PEG3 already some time ago. The principal steps of the processing technique are depicted in Figure 2.16. An anisotropic etch profile results due to the difference in solubility between the exposed and unexposed regions.

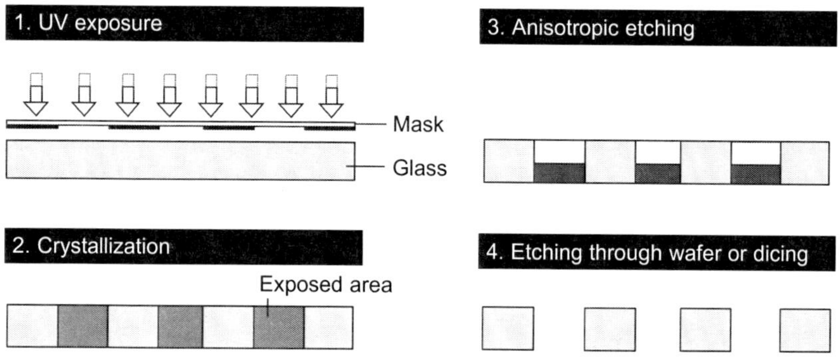

FIGURE 2.16

Photostructurable glass process sequence.

FIGURE 2.17

Microfabricated titer plate made from Foturan®. (Image courtesy and original: mikroglas chemtech GmbH, Mainz, Germany [57].)

An example of photoetched devices is displayed in Figure 2.17. Additional information concerning custom manufacturing in Foturan® can be found on the mikroglas chemtech GmbH website [57]. A new photostructurable glass, FS21, has also been developed at the Technical University Ilmenau, Germany.

2.4.4 Powder Blasting

A method from applications in the macro world adapted uses a stream of abrasive particles. Ceramic-based particles from alumina or silicon carbide are used. Micro-abrasive blasting is an efficient technique for the low-cost patterning of hard, brittle materials such as glass, silicon, tungsten and ceramics. The material is removed from the work piece (target) by the mechanical impact of accelerated particles as depicted in Figure 2.18.

An example of this erosion technique applied to Pyrex® glass is shown in Figure 2.19. This technology has been thoroughly investigated for application to microtechnology. Particles with an average size of 29 μm or 9 μm are used depending on the degree of allowed surface roughness. The smaller

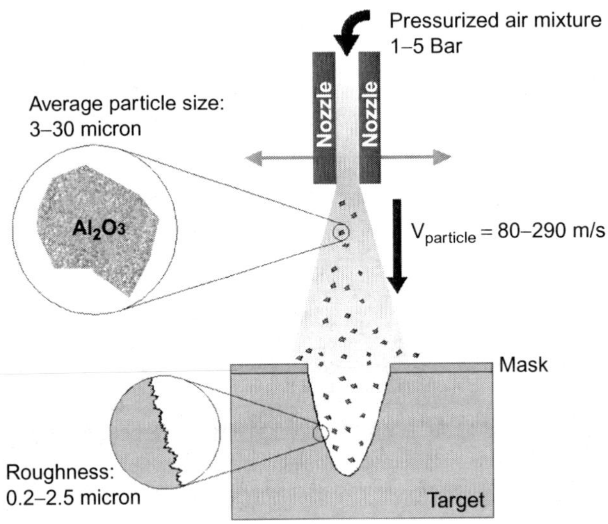

FIGURE 2.18

A schematic diagram of the powder blasting process. (Reproduced from reference [58] with permission from IOP Publishing.)

FIGURE 2.19

Typical result in Pyrex® using powder blasting. Channels are 400 μm deep. (Reproduced from reference [58] with permission from IOP Publishing.)

the particles are, the slower is the blasting process. The definition of process parameters, mask materials and restrictions of applications have been published amongst others in Wensink [59, 60]. Features in the order of tens of microns can be produced, and the technique has been used successfully for producing wafer vias and fluidic microchannels as well as for large-scale production in flat panel displays [61–63]. Although powder blasting is a serial process it is very interesting from the point of rapid prototyping and the manufacture of wafer through-holes in glass micromachining.

2.4.5 Plastic Microfabrication

Plastics are a very special class of materials with a diverse range of properties. Unfortunately, the exact composition or chemical structure of the materials being developed in the polymer industry for microsystems device manufacture has not been specified. An exception to this is the development of polymers for photoresists and as packaging materials for electronics.

In brief, some specific characteristics of plastics materials are: light weight, low cost, large range of tunable optical properties including transparency, low thermal and electrical conductance, and favorable biocompatibility and permeability. A plastic product may also be a polymer blend or composite instead of a single polymer. Plastics suppliers may use different brand names, but are likely to deliver products (monomers, sheets and granulates) of equal quality. This variety in supplies may lead also to additional variables during the design and optimization of a microfabrication process. Within the microfabrication service industry this offers a business niche. A foundry service provider can guarantee that polymer base materials are bought in such large quantities that their properties are consistent, offering their customers one-stop shop solutions for machining and fabrication services, and a source of materials.

Many of the materials used in injection molding or hot embossing could also be used in microsystems. The specific properties of these thermoplastics have to be evaluated and optimized for specific microsystems applications. Thermosetting polymers, such as duroplasts and elastomers, are used more rarely. The silicon elastomer (PDMS) is now being used in micro- and nanofabrication processes by a special technology known as soft-lithography, which is a process that produces a soft (flexible) micro- or nanostructure generated from a hard master. The hard master is usually formed by photolithography.

For plastic micromolding, the number of possible materials is so vast that we will restrict the current discussion to some key polymers and the basic microfabrication techniques that are used. Micro-cutting, -milling and -drilling are also used for shaping polymers. Here, we will focus on four of the innovative polymer-forming techniques that were specifically developed for microfabrication purposes. More detail on these topics can be found in the literature [1, 2, 13, 64]. The examples presented here, are intended as a guideline to polymer microfabrication in the emerging microsystems' market [65].

Thick Resist Lithography

We have already introduced the principle of photolithography for thin polymeric films. These techniques have been taken further through the development of an epoxy-type photoresist called SU-8, which allows single layer lithography up to a thickness of several hundred micrometers. This photoresist is very successful in MEMS because it can be done with an existing UV mask aligner equipment, either directly or with minimal adaptation. Being able to produce structures with such a thickness in standard cleanroom facilities created a kind of microfabrication revolution all around the world. Figure 2.20 presents a patterned layer of 200 μm thick SU-8 on a silicon wafer, illustrating the possibility of vertical profiles [66].

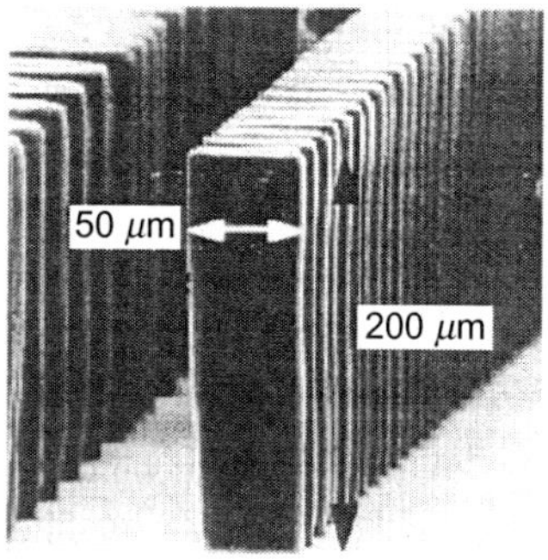

FIGURE 2.20

SU-8 thick photoresist lithography on silicon. (Reproduced from reference [66] with permission from IOP Publishing.)

SU-8 overcame the barriers for patterning three-dimensional structures in mechanical and chemical stable resist. In the early 1990s this was only possible by the use of x-ray lithographic processes. This was difficult because of the need for expensive x-ray beam lines from synchrotron sources. X-ray lithography can be used for patterning structures that require extremely vertical side walls and significant height (very high aspect ratios). Therefore it is often known as a 2.5-dimensional approach in the MST community although complex, three-dimensional features such as photonic crystals or filters can also be generated [67]. The so-called LIGA concept – the German abbreviation of LIthographie (lithography), Galvanik (galvanoforming) and Abformung (molding) – made this process economically feasible. However, it also needed non-standard substrates and has now been replaced by SU-8 in most of the initial application areas. X-ray LIGA is now a niche market and remains as a speciality technique operated by the research groups and institutes that have direct access to a beam line facility. We will discuss LIGA in more detail in Chapter 3.

Realizing three-dimensional microstructures also allowed SU-8 lithography to be used for the fabrication of micromechanical devices (Chapter 5), and molds for galvanoforming and/or direct replication into a polymer (see the sub-section on polymer micromolding below). SU-8 lithography is also popular as a master for the fabrication of mold structures for specific soft-lithography techniques. If patterned SU-8 is used as a mold for galvanoforming processes, it is often referred to as UV-LIGA and can be applied

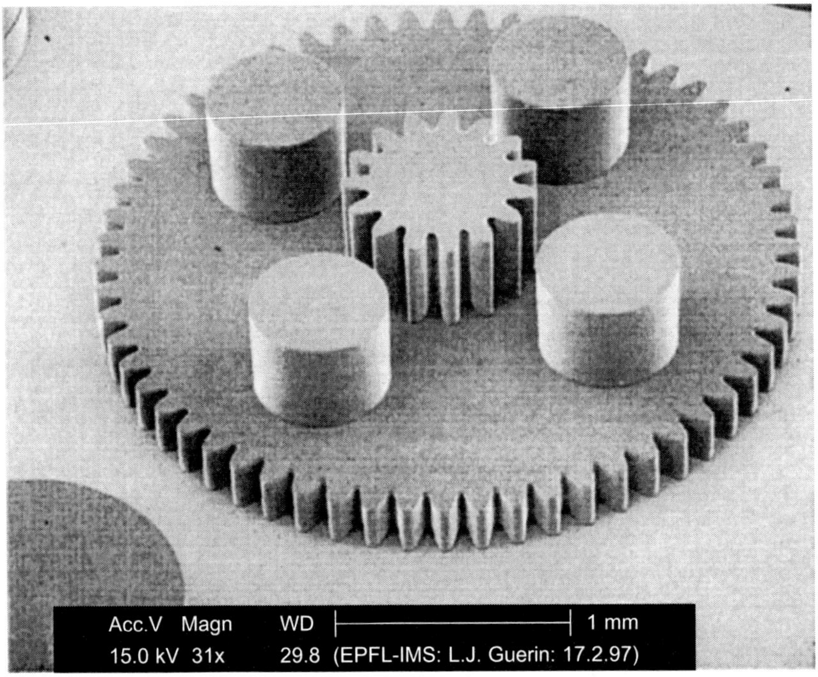

FIGURE 2.21

Multilevel SU-8 master for a watch gear. (Reproduced from reference [68].)

to advanced microtechnologies in combination with replication processes in metal and ceramics. A functional SU-8 photoresist structure may be formed by multilevel UV exposure producing a variety of complex structures, e.g., components of a microgear system such as that depicted in Figure 2.21. The application of this type of microfabrication to miniature gears was successfully introduced to the market by the watch industry.

Laser Ablation

Accurate processing of complex three-dimensional microstructures can be performed by removal of material using excimer or Nd:YAG lasers as an energy source. The laser energy is absorbed at the surface of the work piece and disintegrates the material. Due to the pulsed nature of the process, the heat does not penetrate into the deeper layers of the work piece. Such processing, by superficial removal of material is called ablation. Excimers with

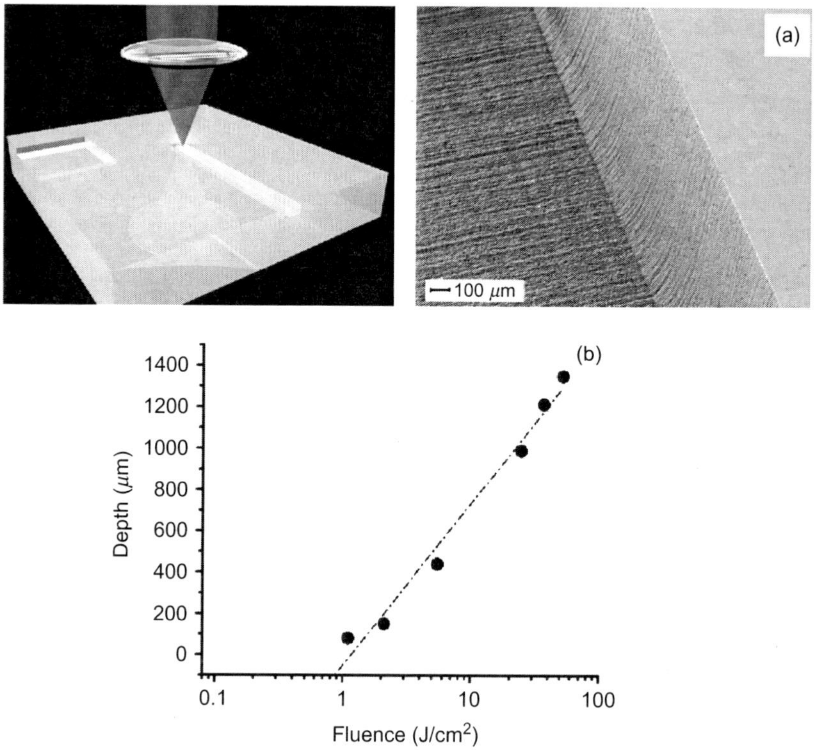

FIGURE 2.22

Top left: Diagram showing femtosecond laser micromachining method. Right: (a) Scanning electron microscope (SEM) image of ablated microchannel produced in polymethylmetacrylate (PMMA); (b) ablated depth versus incident laser fluence. Experimental data points are well represented by a logarithmic (dashed line). The x-intercept of the fitted logarithmic reveals a threshold fluence of 0.85 J/cm^2 (N = 20 pulses per spot). (Reproduced with permission from reference [70]. Copyright ©2010 American Chemical Society.)

wavelengths of 157, 193 and 248 nm are suitable for large area ablation, and Nd:YAG lasers are normally used for small-volume ablation since the latter allow for higher frequency pulses. Figure 2.22 illustrates a laser ablation process and its possibilities for an example of a plastic microstructure. The much longer wavelength of a CO_2-laser may also be used, however, with a less defined pattern accuracy than excimer and Nd:YAG lasers. KrF excimer laser ablation etch rates for polymers such as polysulfone (PSU), polyetheretherketone (PEEK) and polyimide (PI) have been studied as a function of the laser fluence (ϵ). With increased fluence a higher pattern accuracy (inclination angle of the side wall) and reduced debris formation is observed. An overview of various aspects (materials, processes and applications) of the utilization of laser ablation in micro- and nanofabrication processes for polymers is presented by Dyer [69]. Masks may be utilized in excimer laser ablation. Efficiency is increased for complex structures by directly transmitting a large mask geometry in parallel into the polymer surface. The femtosecond laser micromachining principle and an example of a microfluidic channel produced in PMMA is detailed in Marco et al. [70], who specifically investigate the surface roughness of the ablated surface over different laser fluences.

Photopolymerization

Local control of photopolymerization processes initiated by a suitable light source is also considered to be a rapid prototyping technique for complex three-dimensional (3D) polymer microstructures. Obviously price per piece should not be a specific concern when using this technique. However, if only a small number of devices are required, this technique is extremely valuable and effective. A three-dimensional CAD design can be directly translated, layer-by-layer, into the movement of a table containing a photosensitive resin. At the focal point of the laser, the resin gets exposed to such a degree that polymerization of the liquid monomer occurs. The use of a photosensitive polymer that solidifies during exposure has certain drawbacks for the direct manufacture of functional devices or components, specifically with respect to the mechanical strength and hardness of the material. Nevertheless, this process can be used to manufacture maskless polymeric mold structures for subsequent replication processes, either in polymer micromolding (see below) in lost mold replication processes, for example, ceramic microfabrication. The latter will be discussed in Chapter 3.

Microstereolithography is also possible for the direct fabrication of ceramic green-state structures from hybrid ceramic-filled polymer precursors. In Figure 2.23 (left) a schematic of the apparatus is depicted, and on the right of the figure are two types of products. For industrial processes, the current resolution is in the order of 100 μm, and depends on the mechanical accuracy of the elevator displacement of the stage which carries the work piece during the layer-by-layer building process. Currently, attempts

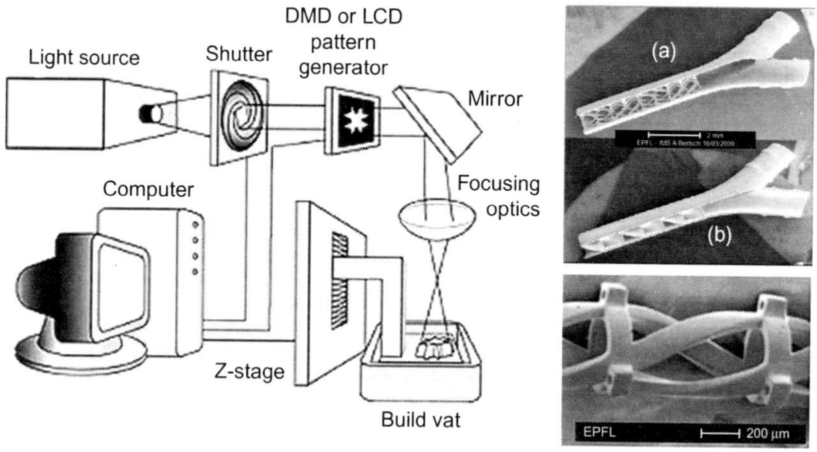

FIGURE 2.23

Left: schematic of a microsterolithography apparatus and typical result thereof. Top right: design studies for micromixers. (Reproduced by permission of The Royal Society of Chemistry [72].) Bottom right: three interwoven springs fabricated by stereolithography. (Figure left and bottom right reproduced from reference [73].)

are being made to reduce this resolution down to the nanoscale by a two-photon photopolymerization approach, in which two beams have to intersect to reach the energy threshold for polymerization. The technique can form 3D objects directly, with a lateral resolution that depends on the focusing spot of the laser beams, but can be as small as 120 nm [71]. UV lasers may also be used for the stereolithographic process, but only infrared (IR) lasers can be applied for the two-photon technique. Since this process is a point-by-point exposure process it is very slow, but it can be speeded up by irradiating the outer boundaries of a design. This process first produces a shell-type structure, which is subsequently solidified, irradiating by UV flood exposure of the enclosed monomer.

Thermoplastic Micromolding

In these micromolding processes, thermoplastic polymers are heated somewhat above their glass transition temperature (Tg). Tg is a characteristic of a given polymer, and one can roughly say that polymers with higher molecular weight also have a higher Tg. Polymethylmetacrylate (PMMA), for example, has a Tg of 150°C, which is a modest process temperature. This allows cycle

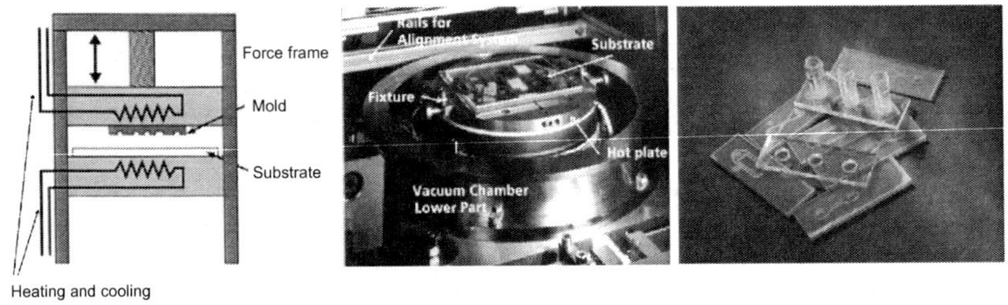

FIGURE 2.24

Left: schematic drawing of micro hot-embossing equipment. (Reproduced from reference [74].) Middle: close-up of a polymer substrate inside the vacuum chamber of a hot-embossing machine. Right: micromolded plastic microfluidic chips for the mixing of pharmaceuticals. (Reproduced from reference [75].)

times of a few minutes to be used for molding by hot embossing. The term nanoimprinting is preferred for the thermally induced re-shaping of a polymer at the nanoscale, which we will further discuss in Chapter 4.

Many different techniques have been demonstrated in the literature, but only a few, such as the fabrication of metal molds by micro-milling or a combination of either x-ray or SU-8 lithography and galvanoforming have to date found their way into industrial-scale MST applications. The machines used for micromolding have also been specially developed. Jenoptik Mikrotechnik, Jena, Germany, for example, originally developed the HEX for applications in micro hot-embossing. Other players also have successfully entered the market using the same thermal polymer printing process at the nanoscale, known as thermal nanoimprinting (see Chapter 4, Section 4.1.6).

Figure 2.24 shows a schematic of a micro hot-embosser, a hot-embossing chamber and resultant microfabricated devices. The examples were supplied by the Microfluidic ChipShop, Jena, Germany, who offer microfabrication foundry services, including micromolding processes.

The mold manufacturing process for micro hot-embossing may be carried out by combining of thick resist lithography (x-ray resist or SU-8) and galvanoforming. Alternatively, it is possible to use micro-cutting, laser ablation, silicon micromachining or other plastic-forming processes (e.g., photolithography) for the construction of a production mold. The mold can be fabricated from one piece, or assembled from different pieces constructed to various accuracy and step height constraints to create the complete mold insert (see, for example, Figure 2.21, Section 2.4.5 above). There are few indicators as

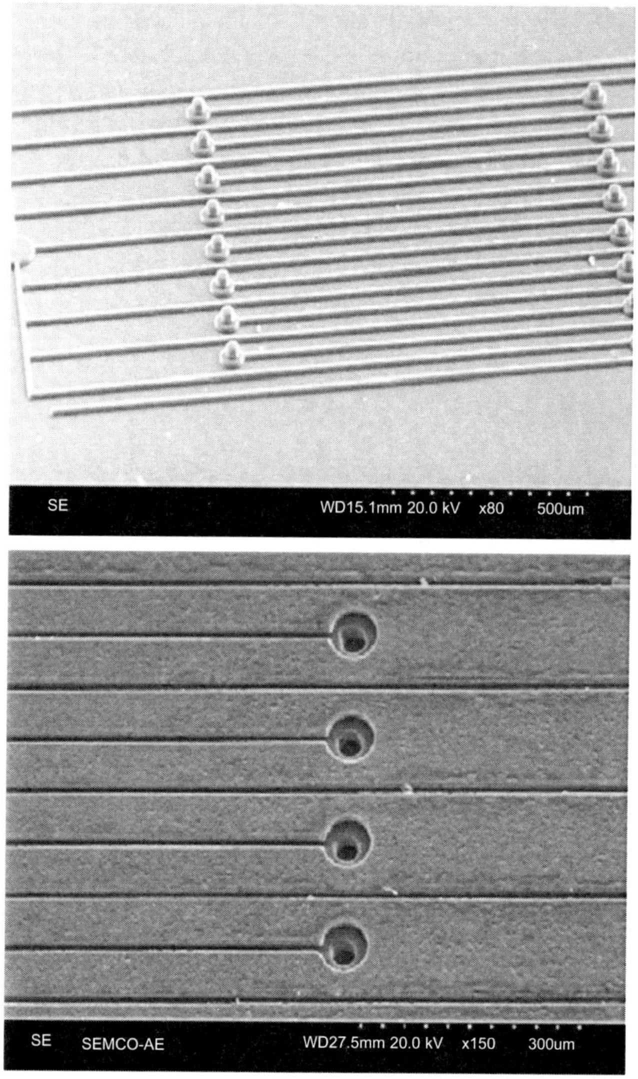

FIGURE 2.25

Printed circuit board (PCB) micro via patterning by thermal imprint technology. (Reproduced from reference [76].)

to which of the techniques is better with respect to a given application challenge. However, metal molds are good temperature conductors, and they are very robust.

Silicon, although mechanically a tough material, is brittle and tends to chip or crack during these harsh applications. For the so-called soft-lithographic molding processes this type of limiting factor is of less concern and these silicon is therefore often used in prototyping and research. Stamps are usually made from nickel, but polymers may also be used to reduce the cost per stamp. However, this type of soft mold can deform when temperatures of several 100°C and high pressure is applied during the mold-filling sequence. The fabrication of micro vias for printed circuit boards (PCB), for example, can

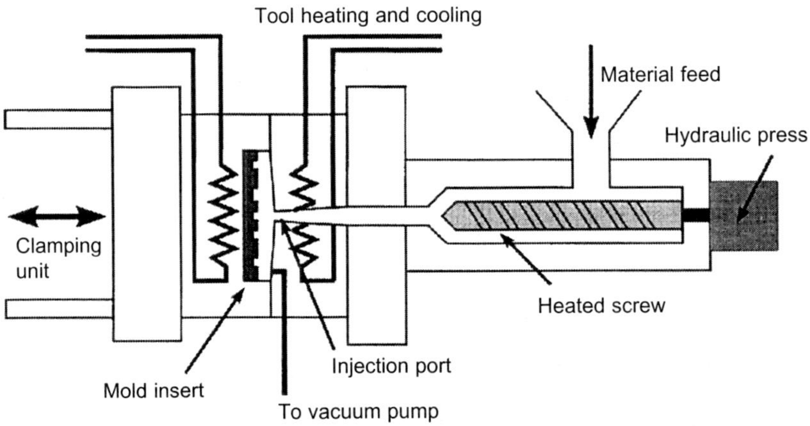

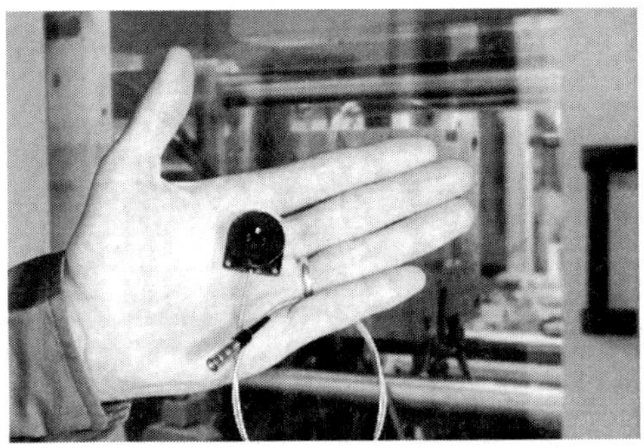

FIGURE 2.26

Top: schematic drawing of an injection molding equipment. (Reproduced from reference [78], Copyright ©2000 John Wiley & Sons, Inc.) Bottom: thinXXS presents a micropump injection molded in Topas®. (Original source: thinXXS, accessed online [79].)

be performed by the imprint technique and is demonstrated in Figure 2.25. The stamp is pressed into a silica reinforced epoxy resin [76].

Besides hot embossing (often now called imprinting for realizing patterns at the nanoscale) injection micromolding is often used when up-scaling to a high-volume production is required. The cycle time, in the order of seconds for injection molding, is the most important parameter in large-scale production. Figure 2.26 depicts a schematic drawing of an extruder feeding the mold tool within an injection apparatus. The lower panel of the figure shows a micro-pump fabricated in the biocompatible polymer Topas® by thinXXS GmbH, Zweibrücken, Germany. Microparts for microfluidic operations are a leading example for the successful introduction of science-based technologies into the market.

Attia et al. have compiled a comprehensive overview of the available literature on microfabrication processes and machines with a dedicated focus on microfluidics structures [77].

2.5 CONCLUSIONS

This chapter has provided the reader with an overview of basic microfabrication techniques. It has covered photolithography and thin-film technologies, adopted from the microelectronic industry, the development of thick resists and other novel material shaping processes. These techniques have lead the way to new applications of microfabrication processes for, e.g., microfluidic devices.

Silicon is an excellent mechanical material and is utilized in diaphragms or special resonant sensors and micro-optical systems involving special comb-type and mass-spring type structures, often requiring high aspect ratio features directionally etched into the bulk of the silicon wafer. The cubic crystallography of silicon offers designers an accurate but relatively simple and low-cost process for achieving a diversity of such microstructures.

Wet chemical etching of other single-crystal materials are also explored for special purpose applications, such as single-crystal quartz in timebase devices.

In many current design processes, plasma-based dry etching is a preferred technology although this choice may be based more on access to equipment than application needs. Surface micromachining for freely moving mechanical structures in silicon has been also developed for their integration with the electronics circuitry on the same platform (silicon wafer). Such monolithic integrated devices are offered as a multi-user process by foundries in the microfabrication service industry.

MST foundry services are increasing in popularity, and more and more academic facilities are opening their doors to industry, not only for research but also for production.

In conclusion, the basic techniques described here reflect the standard processes available nowadays in the microsystems technology industry. The field remains a very dynamic market place, and the continuous development of novel materials will lead to further extension of the microfabrication tool box, some examples of which will be given in the following chapters.

References

[1] M.J. Madou, (Ed.), *Fundamentals of Microfabrication*, CRC Press, New York, ISBN 0-84-939451-1, 1997.

[2] Z. Cui, *Micro- Nanofabrication*, Springer, Berlin, ISBN 7-04-017663-7, 2005.

[3] W.B. Glendinning, J.N. Helbert, (Eds.), *Handbook of VLSI Microlithography*, NOYES Publications, Park Ridge, NJ, USA, ISBN 0-8155-1281-3, 1991.

[4] C.Y. Chang, S.M. Sze, (Eds.), *ULSI Technology*, International ed., McGraw-Hill, New York, ISBN 0-07-114105-7, 1996.

[5] P. Kruit, A. Mohammadi Gheidari, Multi-electron beam systems. In: F. Wakaya, Y. Inoue, J.H. Moon, (Eds.), *Technical Digest of the 22nd International Vacuum Nanoelectronics Conference* (IVNC2009), Art City Hamamatsu, Hamamatsu, Japan, July 20–24, 2009, IEEE, Piscataway, NJ, USA, 2009, pp. 89–90.

[6] J.L. Vossen, W. Kern, *Thin Film Processes II*, Academic Press, New York, ISBN 0-12-728251-3, 1991.

[7] P. Martin, E.M. Skouri, L. Chusseau, C. Alibert, H. Bissessur, Accurate refractive index measurements of doped and undoped InP by grating coupling technique, *Appl. Phys. Lett.* **67** (1995) 881–883.

[8] C. Seassal, J.L. Leclercq, P.J. Viktorovitch, Fabrication of InP-based freestanding microstructures by selective surface micromachining, *J. Micromech. Microeng.* **6** (1996) 261–265.

[9] M. Garrigues, J.L. Leclercq, P. Viktorovitch, III-V semiconductor based MOEMS devices for optical telecommunications, *Microelectron. Eng.* **61–62** (2002) 933–945.

[10] C. Dreyer, M. Bauer, J. Baue, N.J Keil, H. Yao, C. Zawadzki, Polycyanurate ester resins with low loss for use in integrated optics, *First International IEEE Conference on Polymers and Adhesives in Microelectronics and Photonics* (2001) 276–283.

[11] E.M. Yeatman, M.M. Ahmad, O. McCarthy, A Vannucci, P. Gastaldo, D. Barbier, et al., Optical gain in Er-doped SiO-TiO waveguides fabricated by the sol-gel technique, *Opt. Commun.* **164** (1–3) (1999) 19–25.

[12] K. Wörhof, A. Driessen, P.V. Lambeck, L.T.H. Hilderink, P.W.C. Linders, Th.J.A. Popma, Plasma enhanced chemical vapor deposition silicon oxynitride optimized for application in integrated optics, *Sens. Actuators* **A 74** (1–3) (1999) 9–12.

[13] P. Rai-Choudhury, (Ed.), *Handbook of Microlithography, Micromachining, & Microfabrication*, vol. 1, SPIE Optical Engineering Press, Washington, USA, ISBN 0-8529-6906-6, 1997.

[14] Shipley Europe Ltd, *Microposit® S1813 photo resist*. Technical report, Shipley Europe Ltd, Shipley House, Herald Way, Coventry, CV3 2RQ, UK.

[15] T.H. Han, T. Ok, J. Kim, D.O. Shin, H. Ihee, H.-S. Lee, et al., Bionanosphere lithography via hierarchical peptide self-assembly of aromatic triphenylalanine, *Small* **6** (8) (2010) 945–951.

[16] G. Decher, Fuzzy nanoassemblies: toward layered polymeric multicomposites, *Science* **277** (5330) (1997) 1232–1237.

[17] W.A. Pliskin, Comparison of properties of dielectic films deposited by various methods, *J. Vac. Sci. Technol.* **14** (5) (1977) 1064–1081.

[18] W. Kern, Advances in deposition processes for passivation films, *J. Vac. Sci. Technol.* **14** (5) (1977) 1082–1099.

[19] S. Rojas, A. Modelli, W.S. Wu, A. Borghesi, B. Pivac, Properties of silicon dioxide films prepared by low-pressure chemical vapor deposition from tetraethylorthosilicate, *J. Vac. Sci. Technol.* **B 8** (6) (1990) 1177–1184.

[20] M. Nault, T. Weidman, D. Sugiarto, D. Mui, C. Lee, J. Yang, Single layer chemical vapor deposition photoresist for 193 nm deep ultraviolet photolithography, *J. Vac. Sci. Technol. B: Microelectron. Nanometer Struct.* **16** (6) (1998) 3730–3733.

[21] D.R. Cote, S.V. Nguyen, A.K. Stamper, D.S Armbrust, D. Tobben, R.A. Conti, et al., Plasma-assisted chemical vapor deposition of dielectric thin films for ULSI semiconductor circuits, *IBM J. Res. Dev.* **43** (1999) 5–38.

[22] D.H.A. Blank, G. Koster, G.A.J.H.M. Rijnders, E. Van Setten, P. Slycke, H. Rogalla, Epitaxial growth of oxides with pulsed laser interval deposition, *J. Cryst. Growth* **211** (1) (2000) 98–105.

[23] R.A. Buser, N.F. de Rooij, Very high Q-factor resonators in monocrystalline silicon, *Sens. Actuators* **A 21–23** (1990) 323–327.

[24] B.W. Chui, T.D. Stowe, Y.S. Ju, K.E. Goodson, T.W. Kenny, H.J. Mamin, et al., Low-stiffness silicon cantilevers with integrated heaters and piezoresistive sensors for high-density AFM thermomechanical data storage, *J. Microelectromech. Syst.* **7** (1) (1998) 69–78.

[25] K.E. Petersen, Silicon as a mechanical material, *Proc. IEEE* **70** (1982) 420–458.

[26] S. Wolf, R.N. Tauber, *Silicon Processing for the VLSI Era: Process Technology*, vol. 1, Lattice Press, Sunset Beach, CA, USA, ISBN 0-9616721-3-7, 1986.

[27] Y. Cheng, C.-Y. Lin, D.-H. Wei, B. Loechel, G. Gruetzner, Wall profile of thick photoresist generated via contact printing, *IEEE J. Microelectromech. Syst.* **8** (1) (1999).

[28] R. Bischofberger, H. Zimmermann, G. Staufert, Low-cost HARMST process, *Sens. Actuators* **A 61** (1997) 392–399.

[29] K. Fischer, J. Klein, B. Chaudhuri, H. Guckel, Design and fabrication of one millimeter high structures, *Microsystem Technol.* **4** (4) (1998) 176–179.

[30] M. Kufner, S. Kufner, Fabrication of monolithic integrated fiber-lens connector arrays by deep proton irradiation, *Microsystem Technol.* **2** (1996) 114–118.
[31] H.J.G.E. Gardeniers, R. Luttge, E.J.W. Berenschot, M.J. de Boer, S.Y. Yeshurun, M. Hefetz, et al., Silicon micromachined hollow microneedles for transdermal liquid transport, *IEEE J. Microelectromech. Syst.* **12** (6) (2003) 855–862.
[32] Microchemicals. http://www.microchemicals.com/, (accessed 04/04/11).
[33] K.R. Williams, R.S. Muller, Etch rates for micromachining processing, *J. Microelectromech. Syst.* **5** (4) (1996) 256–269.
[34] D.A. Glocker, S.I. Shah, *Handbook of Thin Film Process Technology*, vol. 1, IOP Publishing Ltd, London, UK, ISBN 0-7503-0311-5, 1995.
[35] S.M. Rossnagel, J.J. Cuoma, W.D. Westwood, (Eds.), *Handbook of Plasma Processing Technology*, Noyes Publications/William Andrew Publishing LLC, Park Ridge, NJ, USA, ISBN 0-8155-1220-1, 1990.
[36] R. Legtenberg, H. Jansen, M. de Boer, M. Elwenspoek, Anisotropic reactive ion etching of silicon using SF6/O2/CHF3 gas mixtures, *J. Electrochem. Soc.* **142** (6) (1995) 2020–2028.
[37] M. Mehregany, K.J. Gabriel, W.S.N. Trimmer, Integrated fabrication of polysilicon mechanisms, *IEEE Trans. Electron Devices* **35** (6) (1988) 719–723.
[38] P. Dijkstra, K. Schilp, M. Hausner, Can SEA (semiconductor equipment assessment) also deliver for MEMS? *VIDE* **56** (302) (2001) 684.
[39] B. Schwartz, H. Robbin, Chemical etching of silicon, part I, *J. Electrochem. Soc.* **106** (6) (1959) 505–508.
[40] B. Schwartz, H. Robbin, Chemical etching of silicon, part II, *J. Electrochem. Soc.* **107** (2) (1960) 108–111.
[41] B. Schwartz, H. Robbin, Chemical etching of silicon, part III, *J. Electrochem. Soc.* **108** (4) (1961) 365–372.
[42] G. Schröpfer, M. de Labachelerie, C.R. Tellier, Application and simulation of unconventional bulk-micromachining using underetching of 100 silicon planes, *Microsystem Technol.* **5** (1999) 194–199.
[43] H. Seidel, L. Csepregi, A. Heuberger, et al., Anisotropic etching of crystalline silicon in alkaline-solutions. 1. Orientation dependence and behavior of passivation layers, *J. Electrochem. Soc.* **137** (11) (1990) 3612–3626.
[44] H. Seidel, L. Csepregi, A. Heuberger, et al., Anisotropic etching of crystalline silicon in alkaline-solutions. 1. Influence of dopants, *J. Electrochem. Soc.* **137** (11) (1990) 3626–3632.
[45] K.E. Bean, Anisotropic etching of silicon, *IEEE Trans. Electron Devices* **25** (10) (1978) 1185–1193.
[46] A. Koide, K. Sato, S. Tanaka, Simulation of two-dimensional etch profile of silicon during orientation-dependent anisotropic etching. In: *Micro Electro Mechanical Systems, 1991, MEMS '91, Proceedings. An Investigation of Micro Structures, Sensors, Actuators, Machines and Robots.* IEEE, 30 Jan.–2 Feb., Nara, Japan, 1991, pp. 216–220.
[47] S. Lee, S. Park, D.-I. Cho, A new micromachining technique with (111) silicon, Jpn. *J. Appl. Phys.* **38** (1999) 2699–2703.

[48] G.T.A. Kovacs, N.I. Maluf, K.E. Petersen, Bulk micromachining of silicon, in: *Proceedings of the IEEE, Special Issue Integrated Sensors, Microactuators & Microsystems (MEMS)*, Piscataway, NJ, USA, 1998, pp. 1536–1551.

[49] T. Geßner, E. Vetter, M. Wiemer, Technology tools for a high precision accelerometer in bulk micromechanics, *Microsystem Technol.* **1** (1) (1994) 10–13.

[50] J. Hornsteiner, E. Born, G. Fischerauer, E. Riha, Surface acoustic wave sensors for high-temperature applications, in: *Proceedings of the Annual IEEE International Frequency Control Symposium*, Piscataway, NJ, USA, 1998, pp. 615–620.

[51] L.D. Clayton, E.P. Eernisse, R.W. Ward, R.B. Wiggins, Miniature crystalline quartz electromechanical structures, *Sens. Actuators* **20** (1–2) (1989) 171–177.

[52] N. Sugiyama, D. Yamazaki, T. Ueda, A quartz galvanometer for optical scanning in a laser printer application, *International Conference on Solid-State Sensors and Actuators, 1991. Digest of Technical Papers, TRANSDUCERS '91*, 24–27 June, San Francisco, CA, 1991, pp. 734–737.

[53] E. Momosaki, A brief review of progress in quartz tuning fork resonators, *Proc. IEEE Int. Freq. Control Symp.* (1997) 552–565.

[54] L.B. Koutny, D. Schmalzing, T.A. Taylor, M. Fuchs, Microchip electrophoretic immunoassay for serum cortisol, *Anal. Chem.* **68** (1) (1996) 18–22.

[55] T. R. Dietrich, W. Ehrfeld, M. Lacher, M. Krämer, B. Speit, Fabrication technologies for microsystems utilizing photoetchable glass, *Microelectron. Eng.* **30** (1–4) (1996) 497–504.

[56] K. Yunus, C.B. Marks, A.C. Fisher, D.W.E. Allsopp, T.J. Ryan, R.A.W. Dryfe, et al., Hydrodynamic voltammetry in microreactors: Multiphase flow, *Electrochem. Commun.* **4** (7) (2002) 579–583.

[57] Mikroglas Chemtech GmbH, www.mikroglas.com, (accessed 04.04.11).

[58] H. Wensink, H.V. Jansen, J.W. Berenschot, M.C. Elwenspoek, Mask materials for powder blasting, *J. Micromech. Microeng.* **10** (2) (2000) 175–180.

[59] H. Wensink, J.W. Berenschot, H.V. Jansen, M.C. Elwenspoek, High resolution powder blast micromachining, in: *Proceedings 13th International Workshop on Micro Electro Mechanical Systems (MEMS 2000)*, Miyazaki, Japan, 2000, pp. 769–774.

[60] H. Wensink, *Fabrication of microstructures by powder blasting*, PhD thesis, University of Twente, Enschede, The Netherlands, ISBN 90-365-1698-6, 2002.

[61] D. Solignac, A. Sayah, S. Constantin, R. Freitag, M.A.M. Gijs, Powder blasting as a novel technique for the realisation of capillary electrophoresis chips, *Proc. Eurosensors* **XIV** (2000) 345–346.

[62] S. Schlautmann, H. Wensink, R. Schasfoort, M. Elwenspoek, A. van den Berg, Powder-blasting technology as an alternative tool for microfabrication of capillary electrophoresis chips with integrated conductivity sensors, *J. Micromech. Microeng.* **11** (4) (2001) 386–389.

[63] H.J. Ligthart, P.J. Slikkerveer, F.H. in't Veld, P.H.W. Swinkels, M.H. Zonneveld, Glass and glass machining in ZEUS panels, *Philips J. Res.* **50** (1996) 475–499.

[64] S. Franssila, *Introduction to Microfabrication*, second ed., Wiley, Hoboken, NJ, USA, ISBN 0-470-85-105-8, 2010.

[65] H. Becker, C. Gärtner, Polymer microfabrication technologies for microfluidic systems, *Anal. Bioanal. Chem.* **390** (1) (2008) 89–111.
[66] H. Lorenz, M. Despont, N. Fahrni, N. LaBianca, P. Renaud, P. Vettiger, SU-8: A low-cost negative resist for MEMS, *J. Micromech. Microeng.* **7** (3) (1997) 121–124.
[67] G. Feiertag, W. Ehrfeld, H. Freimuth, H. Kolle, H. Lehr, M. Schmidt, et al., Fabrication of photonic crystals by deep x-ray lithography, *Appl. Phys. Lett.* **71** (11) (1997) 1441–1443.
[68] A. Bertsch, H. Lorenz, P. Renaud, 3D microfabrication by combining microstereolithography and thick resist UV lithography, *Sens. Actuators A Phys.* **73** (1–2) (1999) 14–23.
[69] P.E. Dyer, Excimer laser polymer ablation: Twenty years on, *Appl. Phys. A: Mater. Sci. Process.* **77** (2) (2003) 167–173.
[70] C.D. Marco, S.M. Eaton, R. Suriano, S. Turri, M. Levi, R. Ramponi, et al., Surface properties of femtosecond laser ablated PMMA, *ACS Appl. Mater. Interfaces* **2** (8) (2010) 2377–2384.
[71] T. Tanaka, H.-B. Sun, S. Kawata, Rapid sub-diffraction-limit laser micro/nanoprocessing in a threshold material system, *Appl. Phys. Lett.* **80** (2) (2002) 312.
[72] A. Bertsch, S. Heimgartner, P. Cousseau, P. Renaud, Static micromixers based on large-scale industrial mixer geometry, *Lab Chip Minituarization Chem. Biol.* **1** (1) (2001) 56–60.
[73] A. Bertsch, P. Bernhard, C. Vogt, P. Renaud, Rapid prototyping of small size objects, *Rapid Prototyping J.* **6** (4) (2000) 259–266.
[74] H. Becker, U. Heim, Hot embossing as a method for the fabrication of polymer high aspect ratio structures, *Sens. Actuators A Phys.* **83** (1) (2000) 130–135.
[75] A. Bertsch, P. Bernhard, C. Vogt, P. Renaud, Rapid prototyping of small size objects, *Rapid Prototyping J.* **6** (4) (2000) 259–266.
[76] S. Ra, C. Lee, J. Cho, S. Lee, J. Lee, M. Hong, et al., Micro via and line patterning for PCB using imprint technique, *Curr. Appl. Phys.* **8** (6) (2008) 675–678.
[77] U.M. Attia, S. Marson, J.R. Alcock, Micro-injection moulding of polymer microfluidic devices, *Microfluid. Nanofluidics* **7** (1) (2009) 1–28.
[78] H. Becker, C. Gärtner, Polymer microfabrication methods for microfluidic analytical applications, *Electrophor.* **21** (1) (2000) 12–26.
[79] www.thinxxs.com

CHAPTER

Advanced Microfabrication Methods

3

CHAPTER CONTENTS

The previous two chapters have discussed the philosophy of miniaturization and introduced the basic techniques. We will now further explain and broaden the MST engineering concept. In electronics, the discrete transistor replaced the vacuum tube in the 1960s. Since then, circuitry has been packed more and more densely onto the substrate. In fact, most of the effort in miniaturization was concerned with the development of novel fabrication methods, to allow manufacturing at the microscale of the next generation of devices.

Microfabrication for Industrial Applications. DOI: 10.1016/B978-0-8155-1582-1.00003-4

The reduction of size fitted the immediate purpose of the application: more calculation capacity in the same package. These devices became more portable, due to the reduction in their physical size and weight, and the inclusion of novel means of telecommunication. The development of cordless data transmitting systems was an important step towards the establishment of MST.

This size reduction production has also reduced production costs by fitting more of the same components on one wafer, hence using less material per fabricated device. In industrial microelectronics the lateral feature size has dropped to 90 nm due to improved lithographic methods, manufacturing equipment and processes. Optical lithography approaches today a line width of 65 nm and this may reach 45 nm. However, this is close to the physical limits of the traditional transistor.

While new microfabrication techniques were initially aimed at reducing the size and price of a functional device, micromechanical integration with electronics devices for the sensor industry drove the diversification of technology. Mechanical functionality could be produced by conventional machining techniques, and further miniaturization gave rise to increasingly sensitive and precise sensors and actuator devices.

Many good text books describe these lithographic advances, so, in this chapter, focus will be given to advanced microfabrication methods. We will discuss the processes that allow a broader range of materials to be used at the microscale, and introduce a variety of applications which benefit from these developments, for example, microsystems technology in the optical devices industry.

These novel methods have combined electronics with at least one other discipline, such as optics, mechanics, chemistry and biology, to produce multi-functional systems. Today, in many disciplines, miniaturization is explored from various perspectives. For example, in biology, there is the prospect of developing devices that can interact with cells and proteins.

Generally, one separates research and development in micro- or nanotechnology from the investigation and design of product and process engineering. However, in the modern information society, the potential value of an application drives the efforts to produce it. This means that an extensive set of individual processing steps, often called standard microfabrication techniques, has been built up in this field in a relatively short period of time (three or four decades). Each of these steps is a building block in the design process of a new technology. Utilizing an appropriate sequence of steps to manufacture a specific component or system leads to novel applications and markets, and thus feeds innovation.

In electronics, the processes sequence works on design layers which determine the various deposition and selective removal steps, creating the final device. The architecture of the exact sequence of individual process steps for miniaturization is documented by the designer in a process document, in which each layer refers to a specific mask. During assembly of such process documents, the designers may follow more their intuition or use creativity rather than strict engineering rules. Often new ground is established and novel fabrication working principles are discovered by doing so. Many observations of an operation may originate at the physically restricted nanoscale and are not fully understood. Before going into details of precompetitive science-based technology, however, the following sections will discuss micro-manufacturing techniques which have already been established through advanced process combinations or variations of the basic techniques we have already discussed.

3.1 LIGA

LIGA is a German acronym (LIthographie, Galvanik, Abformung), which refers to the steps of lithography and electroplating used to produce the primary template structure for molding, and producing the replicated secondary microparts, generally formed from a polymer. It is also possible to use photolithography in this process sequence, but the use of highly parallel x-rays as a light source has played an essential role in the development of LIGA as an advanced microfabrication technology. The demand for high aspect-ratio metal structures for the separation of isotopes in the nuclear industry motivated these developments in the early 1980s. The microelectronics industry needed resolution that was not limited by optical diffraction, and this made the beam lines of synchrotron sources more accessible. 1:1 shadow projection using x-rays did not replace the optical lithographic processes in the microelectronic industry, but their use became possible in other applications.

It was discovered that hard x-rays can be used to generate differences in dissolution rate in a high molecular weight polymer such as polymethylmethacrylate (PMMA), while maintaining a very high structural resolution through sheets of several millimeters in thickness. These polymeric structures were subsequently used to generate metal structures at the same structural resolution by electroplating. Special 3D microstructrues, such as photonic crystal-type devices were successfully prototyped by LIGA [1]. Figure 3.1

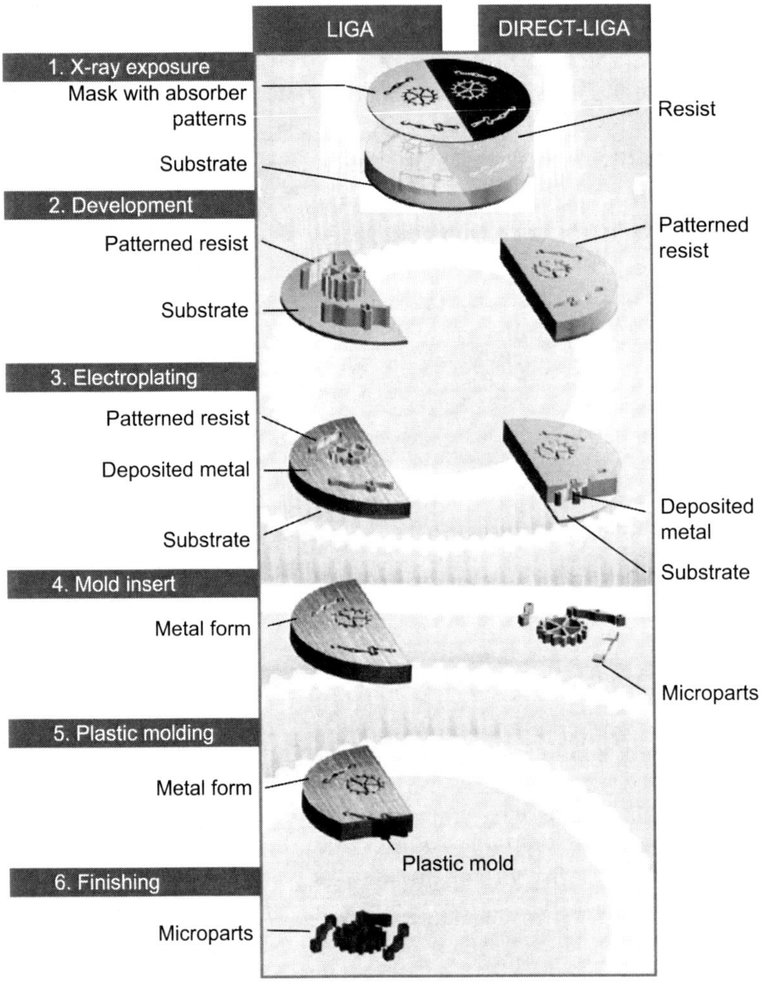

FIGURE 3.1

LIGA process scheme: a means by which a nickel mold can be formed. (Figure accessed online and adapted from reference [2]. Courtesy Helmholtz-Zentrum Berlin.)

depicts fabrication by LIGA using x-ray lithography in the primary patterning process. Microparts can be made directly from the x-ray exposed resist, but it is much more cost-effective to use the original x-ray resist structure as a mold for metal electroplating. This forms a robust metal mold structure and low-cost, tertiary, plastic microparts can then be reproduced from it.

3.2 DEEP REACTIVE ION ETCHING

Chapter 2 introduced conventional parallel plate dry etching. The application of an electrodeless, high density plasma source using electron cyclotron resonance (ECR) was a breakthrough in dry etching technologies [3]. In *Application of Dry Etching to Microsensors, Field Emitters, and Optical Devices*, Stella W. Pang states that a Cl_2 plasma ignited by an ECR source will produce vertical profiles without the need for side wall passivation [4]. However, the selectivity of SiO_2 against Si in the chlorine process is not sufficient. Thin film electrodeposited Ni has a better selectivity but causes stress problems in layers of adequate thickness. Therefore, stress controlled electroplating is performed to increase the mask thickness. Deep reactive ion etching (DRIE), sometimes also called advanced silicon etching (ASE), generally uses inductively coupled plasma systems [5, 6].

Cryo-DRIE is a low temperature process based on fluorine gas that was developed for producing very vertical side walls when anisotropically etching silicon [7, 8]. Optimizing such systems needs increased electron concentrations in the bulk plasma and higher ionization density of reactive species in the dark space (see Chapter 2, Section 2.3.1 for definitions of the mechanisms involved in dry etching). The Bosch process (see Chapter 2) is another example of the use of highly complex plasma equipment in a user-friendly process. The DRIE plasma reactors may be also applied to silica or glass microfabrication. Results of such advanced glass dry etching processes can compete with wet chemical photostructured glass etching (see Chapter 2, Section 2.4.3). Figure 3.2 demonstrates an example in fused silica.

3.3 MICRO-CERAMIC PROCESSING

This section has been inspired by the increased interest in the use of ceramic materials in micro- and nanotechnology. Ceramic processing can employ microfabricated molds, and advanced lithographic techniques which allow volumetric up-scaling of the production of ceramic microstructures, such as the LIGA process, can also be applied to ceramics.

The drying conditions and sintering parameters of a number of ceramics are under investigation for micro- and nanomolding applications. The high temperature resistance, chemical inertness, biocompatibility, conductivity, and/or piezoelectric which is a possible property of ceramics make them potentially valuable in applications. Sub-micrometer lateral resolution has been

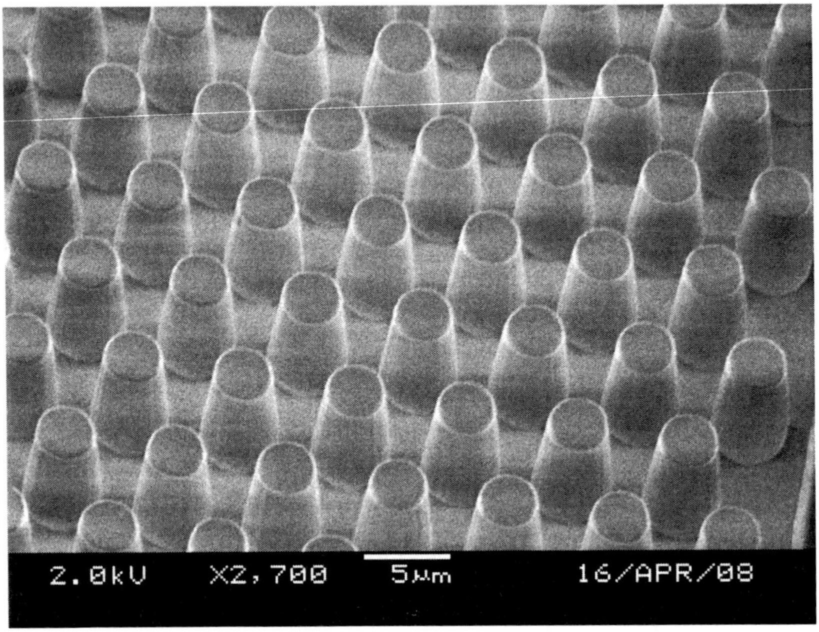

FIGURE 3.2

Inductively coupled reactive ion etching in silica. (Image courtesy: S. Sukas, University of Twente, 2011.)

demonstrated for different types of material systems mainly by depositing the ceramic building blocks directly from the vapor phase, e.g., by sputtering into a hard mold.

Ceramic materials may also by shaped by employing ceramic precursor resins in microsterolithography, which is very similar to the microfabrication of plastic microparts (see Section 2.4.5 in Chapter 2). This section gives an introduction to ceramic micro- and nanomolding as an advanced microfabrication processing technology. Such techniques offer efficient and accurate patterning of these otherwise hard-to-process materials.

3.3.1 Micromolding

Microelectronics has been the driving force behind the developments in microengineering over the last few decades. The thereof emerging fields such as micromechanics, microoptics, microfluidics etc., have reached consolidation phase in research. In industry, especially telecommunication and medical devices markets have stimulated many new developments at the micro- and nanoscale.

Bioceramics, for example, can be used for repair and reconstruction in the body, and have been extensively investigated for their properties as bioinert or bioactive materials, or because of their ability to form porous structures suitable for tissue ingrowth. Potentially, these materials could interact with tissue at much smaller dimensions if their shape, size and ordering were tightly controlled. Micromolding is known for its vast capability in microfabrication of polymer parts, and generally employs a physical template (called also a mold, stamp or master), which is exposed to the liquified polymer. After hardening, the polymer takes the inverse shape of the mold. This technological strategy is particularly popular because of its large-scale production capabilities: the primary mold structure may be expensive to manufacture, but the costs per part reduce dramatically with replication.

The selection of the mold material and accuracy of the features in the mold define the pattern fidelity in the replication process. Mold manufacture, demolding properties and wear-resistance for the production of polymer microparts has been extensively studied. Various types of lithography can be used to make a mold. Lithography employs a resist layer that is normally polymeric, and sensitive to light of a certain wavelength. The pattern is transferred from mask to resist by an exposure step, from a resist to an underlying functional layer by etching, or from a resist to an added layer by electroplating (or other technique) to form a master structure. Many process sequences are known and several books on the topic are available, for example, *Handbook of Microlithography, Micromachining, and Microfabrication* edited by Rai-Choudhury [4]. Photolithography may be applied to manufacture a mold in silicon or to transfer a pattern into a metal by electrodeposition. Other means of mold manufacture are available, e.g., laser ablation, micro-electrodischarge machining (Micro-EDM) or even micro-milling.

Micromolding of ceramics can be done either by utilizing advanced gas-phase deposition techniques to fill the mold with a ceramic precursor material, or by simply pouring a ceramic suspension into the mold. This makes large-scale processing also of ceramic microparts possible. Hard as well as soft molds have been investigated for the molding of ceramic materials. A great many ceramic material systems, including hybrids and composites, are possible and will not be part of this overview.

Nanofabrication in ceramic materials has been much reported in the literature, but only a brief insight is presented here. This discussion is based on a review previously published in the journal *Small* by Auger et al. [9].

Next to silicon micromachining ceramic micromolding could be called a true technology platform, and once it is fully established, many applications may appear. One specific example is the fabrication of microneedle arrays, (see Chapter 8), potentially an ultra-minimally invasive portal to the body for drug delivery and diagnostic purposes.

This chapter introduces ceramic micromolding, demonstrating specifically advanced aspects of this technology.

3.3.2 Ceramic Microparts by LIGA

We have already mentioned that lithographic techniques may be employed for the fabrication of a mold. Although the LIGA process in its original form is not a popular technique, due to the complexity of the light source necessary, the principles involved have often been used for forming a mold for sophisticated, high precision microparts. The tertiary polymer microparts may be filled with metal, or with ceramic fillers. Polymers can be used to provide electrical isolation for the metal structures in a final device, or the polymer structure acts as a lost mold in subsequent high temperature sintering of ceramic microparts. An example of this approach is the use of cements which differ from other ceramic fillers, such as alumina suspensions, in that they set inside the micromold and bond under ambient conditions.

An early example (Colston et al. in 1996), used a LIGA-fabricated nickel mold [10]. This work is influential, since it was the starting point for much materials research. Ordinary Portland Cement (OPC) contains tricalcium silicate as the major component and is used as a model system. The underlying cement chemistry, however, is more complicated in practice; see the original publication for further detail. To improve the final strength, adhesion and permeability of hardened cement, the paste can be modified by polymer additives. The authors of the original paper used styrene butadiene copolymer latex (SBL) as an additional binder between the hydrates and across microcracks and pores. The SBL also contains a surfactant, which enables contact between the hydrophobic polymer mold and the wet cement. Without SBL large cracks develop during hydration causing failure of pattern reproduction. Short hydration times can also increase crack formation. Crack formation during cement micromolding was minimized, and no significant shrinkage was observed when SBL and longer hydration were used. After filling the molds, the cement is cured for up to 16 days and then pyrolyzed within its polymer mold structure from polycarbonate at 400°C for 5 h at ambient pressure. The pyrolysis also burns off the polycarbonate. The polycarbonate lost-mold was formed by the x-ray assisted LIGA method.

Besides demonstrating three different types of cement model systems in this paper, the authors refer only very briefly to the many production variables particle size, choice of latex dispersion, air content, hydrating/curing conditions and parameters of the heating cycle in their concluding remarks. However, these parameters may be better investigated as part of an engineering optimization study, performed if a specific application is considered.

Figure 3.3 illustrates a test feature in OPC after pyrolysis. The structure was hydrated for just 2 days, resulting in a relatively poor reproduction performance, and for 16 days resulting in good replication performance. After long hydration times, the feature contains very detailed, shallow suppressions in the side wall of the ridge structure (originating from the mold made by the LIGA process). The faithful replication of such details in the micro features of a few micro-meters height, illustrates the capabilities of such a process after optimization. A more detailed study can also be found elsewhere [11].

3.3.3 Utilizing Capillaries for Ceramic Micromolding

Micromolding in capillaries (MIMIC) originated from the understanding of microfluidic behaviour in microchannels. Applying the capillary action of microfluidic channels is the key feature of this patterning principle. This discussion centers on its capabilities of patterning ceramic precursors at the microscale, first presented by Kim et al. in 1996 [12]. Figure 3.4 depicts the fabrication process by MIMIC schematically and shows a set of results achieved by this technique with a silica precursor (Accuglass, Allied Signal).

A vast amount of research has been carried out in this field, but little progress has been made in industrial applications. It is possible that limited precursor stability causes defects in the produced microstructures making them not yet fit for market. Perhaps MIMIC is better known as a special form of soft lithography, called microfluidic-assisted lithography, and more publications relating to applications may be associated with this term. Ahn et al. studied infiltration kinetics and drying mechanisms during vacuum-assisted microfluidic lithography of colloidal suspensions and fabricated micropatterned ceramic structures by this method [13]. Optimization of the drying conditions of ceramic films is an important aspect of this type of molding process. See also the work of Guillon et al. [14].

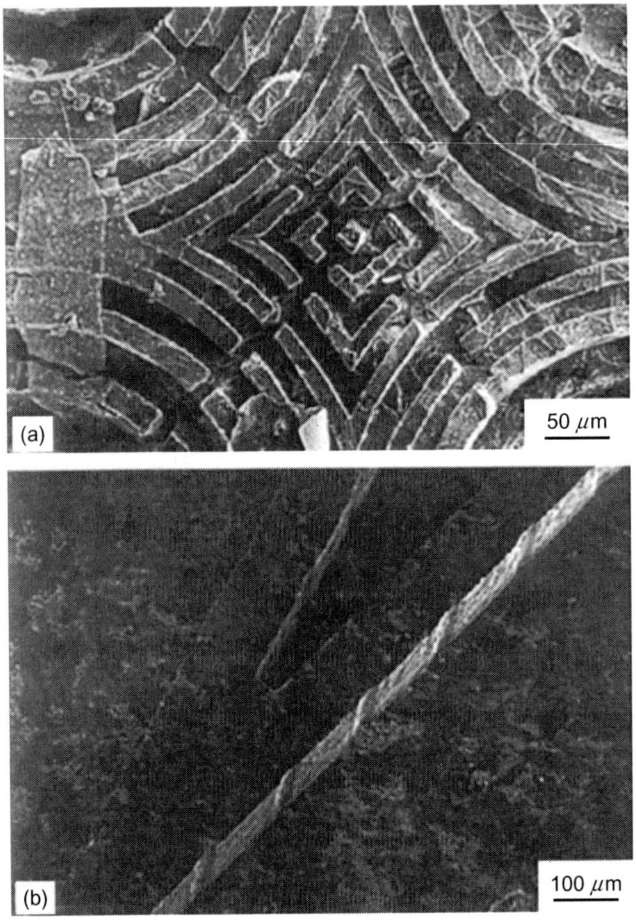

FIGURE 3.3

Pyrolyzed micromolded features in OPC after 2 days of hydration demonstrating limited reproduction performance (a), and detailed feature of the same design after 16 days of hydration (b). (Reproduced with kind permission from reference [10]. Copyright ©1996 Springer Science+Business Media, LLC.)

Silicon carbide (SiC)-based microelectromechanical systems can operate in harsh environments, for example, as actuators in a combustion chamber. Therefore, Lee et al. demonstrated the patterning of SiC by MIMIC and nanoimprinting, and compared the two different fabrication techniques [15]. This publication is a first attempt to compare ceramic properties

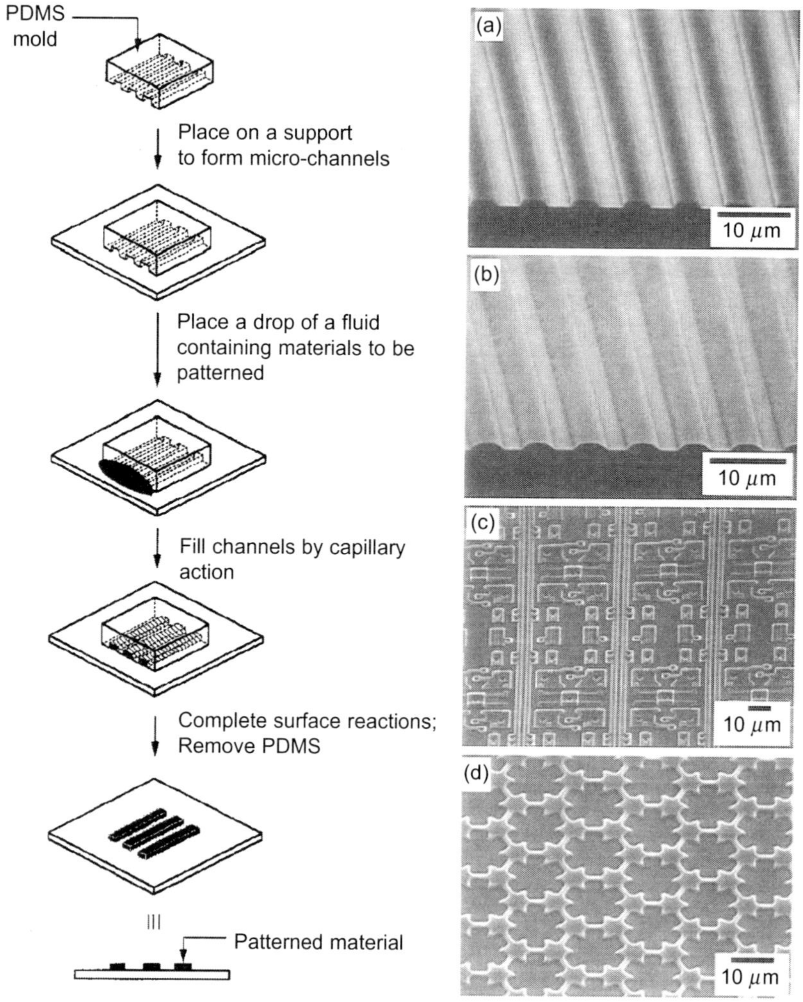

FIGURE 3.4

Micromolding in capillaries using precursor solution to infiltrate the microcapillaries formed by patterned polydimethylsiloxane (PDMS) confocally pressed against a flat substrate (left) and results of this process in silica (right). Patterns of silica/siloxane on glass by partially converting the precursor to silica at 100°C for 20 min. (a); the same structures in (a) were converted to silica at 425°C for 1 h (b); complex pattern of silica/siloxane on Si/SiO_2 (c); and pattern of silica on glass. The precursor film was converted to silica at 425°C for 1 h (d). (Reproduced from reference [12].)

when different fabrication processes are used. The MIMIC process uses a hard-PDMS/PDMS composite as a mold. The mold contained lines and spaces in the sub-micrometer range, and was placed onto a clean silicon substrate to form capillaries by conformal contact between the mold and the substrate. A few drops of the ceramic precursor polyvinylsilazane (Ceraset, polysilazane 20, KiON Corp, USA), mixed with a thermal initiator was applied to the entrance of the capillaries. These are filled by the capillary force which is exerted on the droplets. This liquid precursor was cured for at least 6 h at 70°C in a nitrogen atmosphere. Prior to pyrolysis the PDMS mold was dissolved in 1 M tetrabutylammoniumfluoride (TBAF) in tetrahydrofuran (THF) for 20 min. The cured polymeric microstructures were pyrolyzed at 800°C, resulting in a shrinkage of 19% and 39% in width and depth, respectively, but forming a well preserved sub-micron ceramic pattern, which yielded a film strength of up to 60 GPa.

Microcontact printing utilizes also a soft mold from PDMS, transferring material after spinning or inking the mold structure to a substrate. Nagata et al. demonstrated such a type of ceramic printing process for the fabrication of capacitors utilizing stacked $BaTiO_3$ and conducting $LaNiO_3$ thin films [16].

3.3.4 Utilizing Soft-Mold Replication

An example of this is given by Zhu et al. in their work on forming ceramic microgears from a PDMS soft mold. The mold is generated by replication from a primary microstructure which is fabricated by SU-8 thick-film photolithography [17]. The authors used an aqueous alumina suspension containing 1.5 μm high purity Al_2O_3 particles (Dynamic Ceramic Corp., UK), and Duramex D-3005 (Rohm and Haas, Philadelphia, USA) as a dispersant, and Duramax B-100 and B-1007 were used as binders. The pH of the water-based solution was adjusted by adding NaOH and HCl. The slurry demonstrated best performance at pH 11.

Simple pressurized filling of the microstructures was tested, producing denser ceramic structures. After the green bodies are formed and the binder is removed at 600°C a sinter step with a ramping rate of 200°C/h at 1600°C for 2 h was performed. Figure 3.5 depicts a free-standing microgear with a height and diameter of 1 mm and 2.5 mm, respectively, that was produced in these experiments.

In a subsequent publication, the experiments also used a non-aqueous Al_2O_3/SiC ceramic suspension, mixed with the commercially available

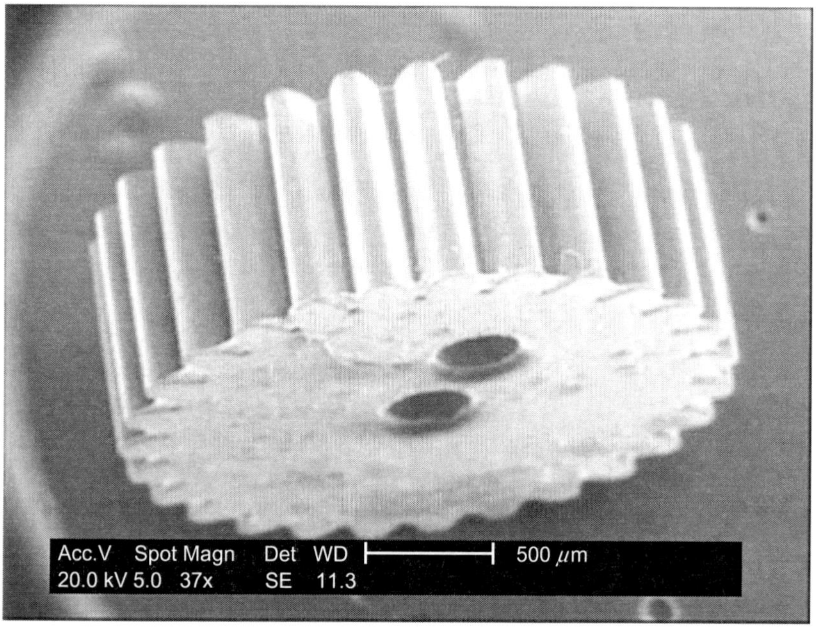

FIGURE 3.5

Scanning electron micrograph of a sintered ceramic microgear. (Reproduced from [17] with permission from IOP Publishing.)

Ceraset Polysilazane 20 (PSZ by KiON Speciality Polymers). This mixture was diluted with acetone (more details of the preparation of the slurry and the curing process can be found in the original publication) [18]. The authors observed that the strong binding force from the cured PSZ prevented both cracking during the drying process, and damage during the demolding step.

3.3.5 Ceramic Patterning on Curved Substrates

This section discusses work by Cannon et al. [19]. The authors first present an extensive overview of the various types of mold being explored for the fabrication of ceramic microparts, and demonstrate ceramic microstructures on curved surfaces for the first time. They also refer to the incorporation of carbon nanotubes in defined sections of the molded ceramic precursor. The combination with nanomaterials is an interesting development, but this text will concentrate on the reported use of flexible PDMS masters to form curved ceramic features which can carry ordered microstructures as small as 20 μm on the inside or outside of a curved surface. These curved surfaces are fabricated by placing the flexible PDMS master with its backing to the curved surface of a macromold (a rod or tube). The micropatterned side is then exposed to the ceramic precursor. The authors used a commercially available low-shrinkage precursor (Cercanam by Ceramatec Inc.) with a composition of greater than 95% alumina with the remaining 5% percent being a proprietary formula. The slurry can be molded or casted into molds with a feature size as small as 5 μm. First the PDMS master is placed onto the curved surfaces with a 3 mm curvature radius. Subsequently, the ceramic precursor is poured onto the molds, and cured by heating at 75°C for 2 h to harden the precursor into a green body. At this stage the green body can be released from the mold, and be cured at 450°C for an additional 2 h. Figure 3.6 illustrates this process. The mold is mounted such that the microstructure is positioned at the outside of the curved surface (Figure 3.6(a) step B). The subsequent replica is therefore inverted, and carries the microstructures in the inside of its curved surface (Figure 3.6(b)), while the details of the features at the microscale are preserved (Figure 3.6(c)).

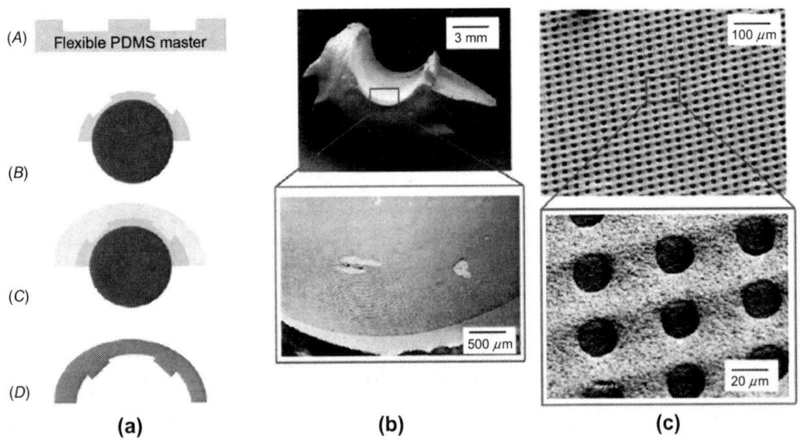

FIGURE 3.6

Schematic fabrication process steps A–D (a), and resulting microstructures molded inside ceramic with a curved surface (b) as well as details of the resulting microstructure (c). (Reproduced and rearranged from reference [19] with permission from IOP Publishing.)

The ability to micromold ceramics onto a curved surface may impact on the fabrication of complex metal structures by the so-called investment casting method. Flexible PDMS microstructures are transferred into a ceramic structure containing curved as well as microstructured elements. The ceramic part is then filled with, for example, molten aluminum with a melting point of 695°C. The ceramic form is consequently destroyed to form the metal structure, which may then be used as a mold for the production of a large number of polymeric microparts [20].

3.3.6 Patterning Ceramic Materials at Nanoscale Resolution

During the mid 1990s the micromachining community explored many lithographic techniques, to attempt to further reduce the physical dimensions of structures, and so pack more and more functional devices onto an electronic chip with the same area, typically 36 mm × 36 mm. These attempts included self-assembly processes, which involve the deposition of thin films, molecular monolayers, block co-polymers or sub-micrometer sized particles onto a template. The template constrains the position and orientation of the grains or particles, creating highly-ordered features by physically binding the building blocks at pinning points. These pinning points are designed in the template to create hierarchically ordered layers.

The method also provides a certain level of registration for these extremely small features. In some papers this process is called graphoepitaxy, as it is based on the directional solidification of block co-polymer films. Various templating techniques were investigated to produce nanometer-scale parallel patterning of ceramic and organic–inorganic hybrid materials, often based on the soft-lithography approach. A comprehensive review by ten Elshof et al. has summarized these investigations [21].

Auger et al. give an introduction to physical vapor deposition techniques that have been used with templates at the nanometer scale [9]. In this approach the ceramic material itself is directly "molded" (constrained by the template during film growth) instead of its precursors, by using either soft or hard molds, and avoiding the need for post-patterning treatments. To do this the molds must have good anti-sticking properties, achieved by special coatings utilizing so-called surface self-assembled monolayers (SAMs). Silicon molds, for example, have been surface modified by immersion in the SAM precursor, octadecyltrichlorisilane (OTS). Metal molds may be primed by alkanethiolate SAMs to give good anti-sticking properties.

The temperature of the substrate must be controlled during vapor deposition, to ensure that the SAM remains chemisorbed on the surface of the mold. The sensitive SAM layers can be protected by increasing the substrate-target distance and utilization of an inert gas in the sputter chamber. A specific challenge for ceramics is their brittleness, which complicates the demolding process. The thin ceramic films must therefore be backed, for example, by using Scotch® tape or, in a more sophisticated process, by bonding the ceramic film to a glass handling wafer before the demolding step. Free-standing features are therefore difficult to achieve. Auger et al. demonstrated the pattern transfer from a mold containing nanodots for the patterning of AlN, TiN and B_4C films [9]. Figure 3.7 illustrates an example of such nanomolded structures.

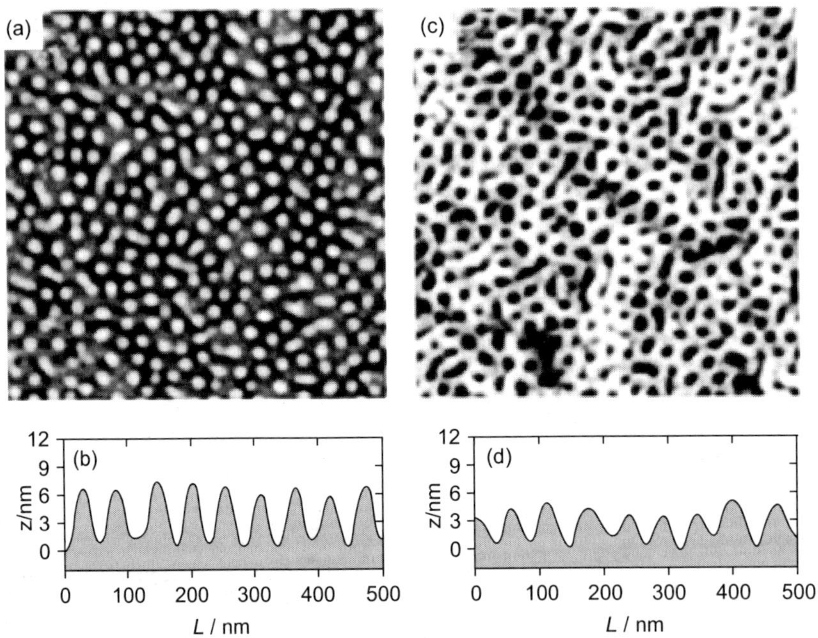

FIGURE 3.7

Nanostructured ceramic material patterned by replication. Top-view atomic force microscopy images (1 x 1 μm^2) and cross-sectional analysis of the original nanostructures in SiO_2/Si (left) and the corresponding nanostructured TiN surface (right). (Reproduced from reference [9]. Copyright ©2000 John Wiley & Sons, Inc.)

3.4 SPECIALITY SUBSTRATES

3.4.1 Silicon-on-Insulator (SOI)

Silicon wafers have been used abundantly in microelectronics and MEMS as a platform for fabrication. An interesting variation of the standard silicon wafer is the silicon-on-insulator substrate. To produce these wafers two silicon wafers are bonded together, by using silicon dioxide of approximately 1–2 μm thickness as a bond layer. One of the silicon wafers is thinned down to a thickness of 10–50 μm. The exact layer thickness will depend on the application. Wafer thinning is performed by wet etching so this preparation method is often called the *bonding-and-etchback* technique. The buried oxide layer (BOX) is not accessible unless bulk micromachining is performed, either on the thinned top or full thickness silicon wafer, which also acts as a handling platform for processing. Alternatively, a special cracking technique involving hydrogen implantation, bonding and heating of the wafer stack to 470°C can be used to create a thin silicon wafer on top of the BOX layer. Bonding of a 300 μm quartz wafer sandwiched between two standard 525 μm thick silicon wafers was also demonstrated [22].

These speciality substrates find applications in power electronics, electronic operations for application at elevated temperatures, bulk micromachined xyz-stages, radio frequency (RF-MEMS) switches, microheater devices for microfluidic applications and integrated optical systems.

3.4.2 Electro-Optic Substrates

Another group of speciality substrates are materials that carry the electro-optical effect, for example, $LiNbO_3$ [23]. These types of materials are important for the design of so-called surface acoustic wave (SAW) devices. In the ferroelectric $LiNbO_3$ and other SAW device materials SAWs are excited by interdigitated metallic finger electrodes, which are fabricated by depositing a thin metal layer (gold, aluminum) on the substrate surface and lithography. These devices form electronic filter and signal delay lines and are mainly manufactured for television and mobile communication applications. SAW devices therefore represent a huge market.

3.5 ADVANCED NON-SILICON AND SILICON HYBRID DEVICES

The selection of a specific material at the component level is based on its unique properties (e.g., optical, piezoelectric, magnetic, super-strong, lightweight, self-cleaning, shape-memory and other smart properties). Integration of these materials with other integrated components, specifically electronic devices, can be highly challenging. One of the preferred solutions is to process the components in different production lines, and merge them at some stage of back-end processing to generate a hybrid integrated system. The following two subsections will give brief examples of such approaches.

3.5.1 Nanofabrication of Information Storage Devices

One of the early adaptors of nanotechnology is the data storage media market. Technical developments in the fields of optical and magnetic data storage media have been adopted at an extremely high pace. Research findings from as recently as 10 years ago are currently being tested for application and there is no doubt that this industry will grow further.

Nanotechnology plays an evermore important role in information storage. The need in the market for functional elements that approach the physical limit of so-called patterned media that have dot dimensions and pitches of the order of a few nanometres is an enormous driver for research and development in this area. Below this limit, thermal noise dominates and the distinguishable writing and reading of a single dot becomes impossible. At the same time research continues to explore other means of information storage, e.g., in the capture and release of energy through chemical bonds in single molecules. We will discuss examples of the fabrication of patterned magnetic media in Chapter 4, which covers nanofabrication.

3.5.2 Integrated Optics

Optical devices and systems, such as CD-ROMs, DVDs, and TFT image displays, are part of our daily lives, both at home and on the move. A vast amount of information can be easily accessed anywhere via the World Wide Web, by devices such as PCs and mobile telephones. The business impact of this type of development was cited in market research reports that claimed extremely high growth rates for the fibre-integrated market sector between 1999 to 2003 due to the huge increase in commercial and public use of the internet [24].

This technology has made an enormous impact on our lives regardless of its commercial success. Also we now know that economic welfare and flourishing, sustainable businesses are probably more closely linked to a stable banking system than sole technical innovation.

The introduction of micro components has revolutionized the entire field of photonics. Optical phenomena depend on the careful alignment of all system components, which could not be realized in traditional bulk free space optics. Because free space optic components are usually mounted on optical benches, system operation degrades due to ambient variations, which still demands further research and development.

New research has, however, resulted in the idea of micro-optical benches, so converting the optical system into a compact, stable (vibration tolerant) package [25–27]. Early work in photonic device development followed the same approach as planar thin film technologies, producing opto-electronics components. But the need soon arose for optical modulation, beam steering and input/output coupling of three-dimensional, actuated micro-optical switch devices, such as the one depicted in Figure 3.8 [28]. This category of device belongs to the novel field of photonics devices that was founded on developments in MST/MEMS technology, and so they are named micro-opto-electromechanical systems (MOEMS).

Another example in this field which clearly demonstrates the technology gap between scientific research and final development is the semiconductor laser. Although predicted in the late 1950s and developed around 1960,

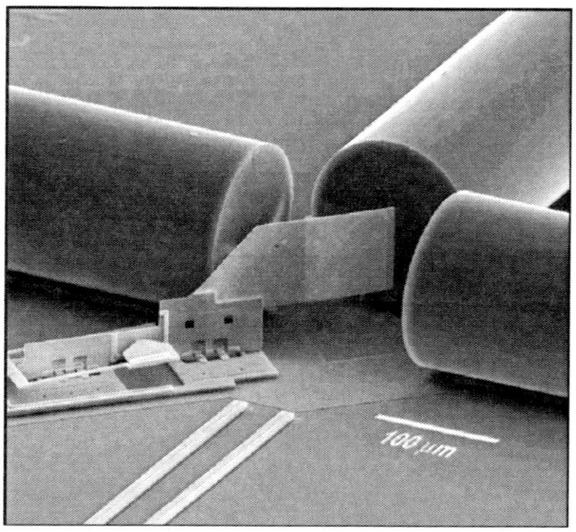

FIGURE 3.8

Integrated micro-optical switch device. (Reproduced from reference [28]. Copyright ©2001 Materials Research Society and Cambridge University Press.)

the commercialization of laser diodes did not start before the early 1980s [29]. It appears equally difficult to bridge this technology gap in MEMS developments, although some preliminary commercial success was achieved during the dot-com era. Economic losses in this period are in no way related to a lack of technical possibilities, but investments could have been more selective at that time.

On the other hand thin-film related optical integration has steadily increased over recent years, and is clearly currently better suited to industrial applications than are free space micro-optics. The next section will give examples of important developments in the field of planar integrated optics.

3.6 PLANAR LIGHTWAVE CIRCUITS

A difference in optical refractive index between two materials can be used to guide photons. The optical fibre cable is the most efficient long distance information carrier that we currently know, with transmissions reaching an optical loss of less than 0.4 dB/km at 1.3 μm [30]. The advantage of a circular fibre is its extremely sharp transition from the core to the cladding. Pulling the core from a master means that the core is embedded into the cladding with a very smooth surface [31]. This degree of side wall smoothness and radial symmetry is difficult to achieve using planar patterning methods. The higher the roughness of the encapsulated surfaces, the higher is the optical power loss. When fabricating planar optical circuits, similar requirements to the fibre cable are expected. A highly optically homogeneous material must be shaped into a guiding device. For slab waveguides, this problem is to do with the surface quality of the substrate and the grain properties (morphology) of the deposited layer. However, in channel guides, optical transmission is also restricted by losses through the side walls. This roughness must be reduced to produce a device of the required standard, and to do this, an appropriate technique used to pattern the waveguide. Patterning methods can be divided into three groups:

1. subtractive pattern transfer through a masking layer,
2. additive processing by lift-off, or
3. selective change of refractive index, for example, by doping.

Refractive index modulation relies on either diffusion at high temperature, ion exchange or high energy ion implantation. Very sophisticated features can be made by pulsed laser techniques, which may be also suitable for selectively changing material properties and therefore being suitable for direct writing of waveguide devices [32, 33]. Refractive index control and the transition towards the cladding layer are non-trivial in all of the three techniques listed above. The easiest approach is the deposition of a highly homogeneous film of defined thickness and refractive index onto a substrate and then etching away the excess material to create the guide structure.

The optically guiding layer on the substrate is called a slab waveguide, and they are the building blocks of optical devices which are used to confine information in the optical frequency domain and transfer the energy traveling along the surface-defined optical layer. Optical tracks can be embedded in the substrate by splitting and merging these guiding tracks in a similar way to electronic circuitry. Hence these devices are referred to as planar lightwave circuits (PLCs). The optical material may be deposited in layers from solution, applied by spin-coating, or by dielectric film deposition techniques such as flame hydrolysis, chemical vapor deposition or sputtering. However, to achieve a 3D channel, the lateral dimensions of the material must be defined, too. This waveguiding effect is sensitive to roughness as discussed above. Scattering on a rough surface creates optical loss and inhibits the achievement of high quality guiding properties. Optical loss is also caused by curved side wall profiles.

Wet chemical etching has only limited use for producing channel waveguides. The high specifications of photonics call for specific fabrication processes in suitable materials and with high pattern fidelity with respect to the optical functionality. Other applicable techniques are doping by diffusion, ion beam implantation or ion exchange, but the resolution of these techniques is limited [30, 34]. Dry etch processes can compensate for the disadvantages of wet etching and produce high resolution structures. Selecting the exact sequence of the process steps requires a detailed knowledge of material properties, and hence must be altered for each material.

Although novel photonic crystal materials have been introduced, most industrial devices are made by combinations of bulk and surface silicon micromachining as well as hybrid assembly of active optical and waveguiding components by lithography. The lift-off process could also be capable of attaining the required side wall quality if precise lithographic control and very thin films are used. Use of lift-off techniques with polymer masks does, however, restrict the process temperature to around 200°C. In light of these arguments, the removal of the unwanted materials via dry etching appears to be the most feasible approach.

Many waveguide processing protocols integrated with micromechanical devices have been developed in university-based research groups. For example, at the Optical and Semiconductor Devices Group, Imperial College, London, UK, an in-house RIE process for silicon dioxide etching has been developed, which consists the following steps:

1. Thermal dry oxidation of the silicon wafer at 1100°C for 24 h;
2. Sputtering of a Cr-layer to act as mask material;
3. Cr-mask patterning by photolithography and wet etching in a standard Cr-etchant;
4. Resist removal;
5. Silica dry etching in $CHF_3/Ar/O_2$ gas mixture through Cr-mask, etch rate: 50 nm/min (experimental result for Plasma Technologies RIE 80 equipment).

In general, surface micromachining is considered important to the development of micro-opto-mechanical integrated sensor devices and is often described in the literature, specifically with respect to the patterning of waveguides [35, 36]. Wet and dry processes have advantages and disadvantages for making channel waveguides. The processing techniques used in integrated optics are so similar to the processes of very-large-scale integration (VLSI) microelectronic techniques that they will not be described in detail here. Instead, we give an overview of optical specifications for waveguides and their manufacturing techniques, either applied or considered in the design methodology of sensors and actuators. Fundamental information about the fabrication of integrated planar optical devices can be found elsewhere [30, 37, 38].

TiO_2, Ta_2O_5, Al_2O_3, Si_3N_4 and SiO_xN_y are some of the most common materials for visible light (633 nm) waveguides on glass or silica-on-silicon substrates. To minimize the optical power required to illuminate the target and retrieve a signal from the scattered light in the case of scanner devices, the loss of the waveguide structure should be made as low as possible. This optimization can be addressed by an index difference between core and cladding (or buffer) of at least greater than 1% (specifically for silicon-based optical integration a higher degree of index difference is preferred). The overall length of a straight waveguide will determine the attenuation allowed (for integrated sensing devices, a length of 15–25 mm is probably sufficient), i.e. higher levels of loss may be accepted for these relatively short guides. The surface and side wall roughness should be as small as possible, e.g., a roughness value as low as 5 nm rms could already convert 10% of the fundamental mode into higher mode energy, which cannot propagate along the core for more than one meter, although the effect is wavelength dependent [31]. The roughness/shape of the end face, which is often used for coupling to the light source, determines the coupling loss. If a waveguide is used to illuminate a sample, the roughness/shape of the waveguide ending will determine illumination properties, such as resolution, intensity and stray light loss.

Finally, the intrinsic microstructure of the material itself is important. Layers with small grain size and low porosity will reduce the attenuation of the propagated waves. The choice of a substrate material is also important. Silicon, for example, has a very high refractive index (3.4). To provide a guide for light in the visible range, the waveguide core has to be isolated from the substrate to avoid leakage of energy. Planar silica-on-silicon light circuits are made possible by applying fairly thick optical buffers, which prevent guided modes from coupling into the substrate. This allows the realization of low-loss waveguides on silicon. Figure 3.9 compares the refractive indices of differently deposited materials with the index of selected substrates.

The optical buffer layer between the waveguide core and the substrate must have an appropriate thickness. Figure 3.10 shows the relationship between buffer layer thickness and the thickness of the waveguide layer as it relates to the refractive index of the guiding material [37]. The graph demonstrates that for single mode light propagation with a layer thickness of approximately 0.3 μm and a refractive index of 1.7 at $\lambda = 633$ nm, the thickness of the buffer must exceed 1 μm.

In most cases an amorphous glass film is a suitable material for optical waveguides in the visible wavelength range. The amorphous films are deposited onto glass substrates, which are also amorphous (e.g., fused silica and thermally grown silica buffer layers on silicon wafers) [40]. In general, any substrate with a smaller refractive index than the guiding layer should provide some degree of light transmission. In addition, single-crystal films can be grown via epitaxial techniques onto a single-crystalline substrate, if the lattice constants of both the film and the substrate are closely matched.

Epitaxial growth is widely applied in optoelectronic device fabrication. For example, Kim and Yoon have studied ZnO/InP heterostructures [41]. The work describes the epitaxial film growth of ZnO on InP substrates using low temperature rf-magnetron sputtering, which is of considerable interest for optoelectronic devices. The paper reports on the material properties characterized by atomic force microscopy (AFM), Auger electron spectroscopy

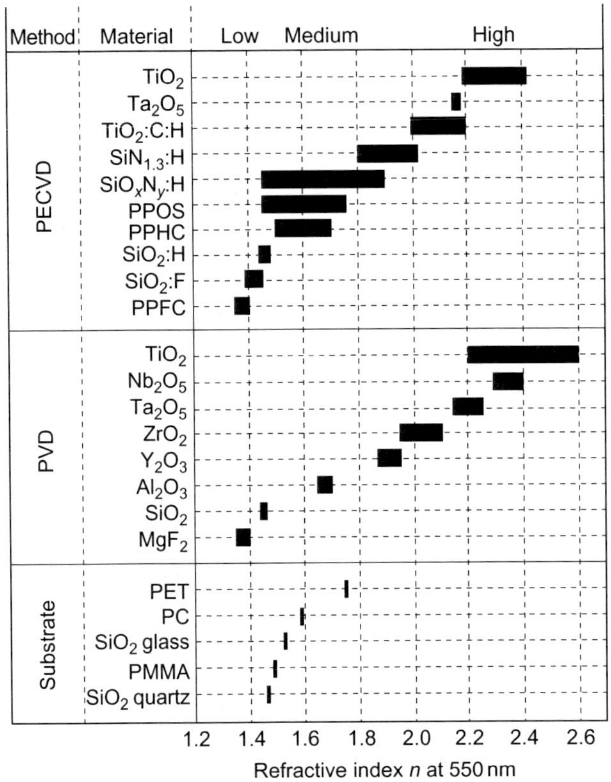

FIGURE 3.9

Refractive indices (at $\lambda = 550$ nm) of optical thin films. (Reproduced from reference [39]. Copyright ©2000 American Vacuum Society.)

(AES) and transmission electron microscopy (TEM). The equipment used is not described further, but a review of optical film technology can be found in Martinu and Poitras [39].

Polymers can also be used as waveguiding layers. Organic materials are, however, more difficult to incorporate into a post-processed substrate. Such a substrate might carry fragile electromechanical features and therefore place higher demands on the polymer coating technique than a planar substrate. Dip coating and plasma polymerization are alternatives to the commonly applied spin-coating technique. The simple application of polymers,

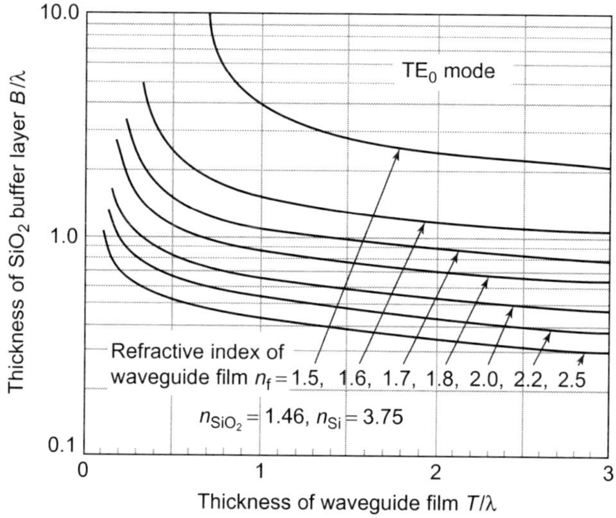

FIGURE 3.10

Computed data on buffer layer thickness. (Reproduced from reference [37]. Copyright ©1989 The McGraw-Hill Companies, Inc.)

via spin-on methods on bare substrates, have produced remarkable results, but this approach has not yet been widely used for applications [42–44].

Silica-on-silicon has produced good quality planar waveguide circuits and can be produced prior to bulk-etch processing [45]. Materials such as silicon nitride are well known from passivation or masking layers in microtechnology and are highly stable, hence they have been thoroughly investigated as guiding layers because they allow refractive index tuning over a wide range (see Figure 3.9). They can also be deposited in very short periods of time (a few minutes for PECVD) compared to sputtered metal oxides, which are deposited at a much lower pressure region and require pumping times of up to several hours, although material deposition rates may be comparable to CVD processes.

Deposition technology at low temperature has been also developed by applying a photo-CVD process. Such photo-assisted deposition uses xenon or krypton chloride excimer lamps to provide energy for the chemical reaction [46]. TiO_2 as well as $(Ta_2O_5)1\text{-}x(TiO_2)x$ on silicon has been reported [47, 48]. Table 3.1 compares the characteristics and film properties of the various processes.

Table 3.1 Optical thin films and their deposition processes

Ref	Substrate	Film	Process	T (°C)	Thickness	rms	Refractive index at 633 nm
[49]	Si-SiO_2 (0.5 μm)	LiTaO	Sol-gel spinning/ annealing	600	0.81 μm	6.3 nm	1.84
[41]	p-InP (100)	ZnO	rf-magnetron sputtering in Ar	200–500	200 nm	1.72 nm	1.93–2.0
[50]	Si-SiO_2 (3 μm)	CNx:H	PECVD	250	320 nm (5% accuracy)		Positive electrode 1.6–1.8; negative electrode 2.2–2.6
[51]	Si-SiO_2 (2 μm)	Al_2O_3	Magnetron sputtering, annealing	800			1.65–1.70
[48]	Si	TiO_2	photo-CVD	<400	<140 nm		2.4 (λ unknown)
[52]	Si	SiO_2 doped with Ge	microwave ECR PECVD	<120			
[53]	Si	SiON	PECVD	<350	300 nm		1.7

3.7 FABRICATION EXAMPLE OF AN INTEGRATED OPTICAL DEVICE

The aforementioned techniques for processing waveguides are currently widely available for their integration into sensor devices. This section presents the application of these fabrication methods to integrating optical waveguide sensors with an active optical function. Both light generation and light detection rely on semiconductors. As mentioned previously, the pure transmission of light can be realized with passive materials, such as amorphous glasses.

Dry etching in a gaseous atmosphere at low pressure is the method of choice for the industrial production of waveguide features. This process can be integrated into standard microelectronic circuit manufacture as post- or pre-process to the so-called complementary metal oxide semiconductor (CMOS) technology. Besides the higher accuracy that is achieved by dry etching, it is also compatible with CMOS, which often only allows post-process integration.

Several types of dry patterning equipment are available. Sputter systems are the oldest; originally operated in a back-etch mode to clean the substrate surface prior to thin-film deposition. Most processes for optical waveguide etching currently use a dedicated reactive ion etcher (RIE), as described in Chapter 2 for silicon etching. For reactive ion etching of waveguides, the gas must be carefully chosen to yield high pattern fidelity at a high etch rate (selectivity) for the waveguide material against the optical buffer layer.

Section 3.6 introduced a wide variety of materials used for optics integration. Etch parameters are very dependent on the exact composition of the material and its film morphology as well as machine conditions. Often deposits at the reactor wall can influence and disturb subsequent processes. One example of a specific waveguide design is the integrated Mach–Zehnder interferometer depicted in Figure 3.11. These interferometers have been demonstrated in a variety of applications, such as modulators, couplers and sensors. They can be also integrated with micromechanical sensors to read a mechanical measurement signal [54, 55].

Figure 3.11 shows the schematic principle of the waveguide design and the etched channel waveguide achieved in a Germanium-doped silica layer on top of a silica buffer layer of 20 μm thickness. The integrated optical device is fabricated using the process sequence given in detail in Figure 3.12. This type of device contains critical curved elements that not only require smooth

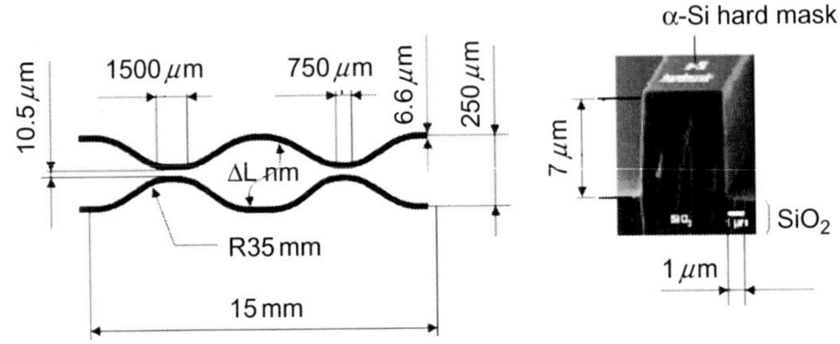

FIGURE 3.11

Wavelength and polarization insensitive integrated directional coupler using Mach–Zehnder structures. Parameters used to model the device performance are: $\Delta n/n = 0.3\%$, $\lambda = 1550$ nm, polarization TE (transverse electric). (Figure redrawn from reference [56], with permission by Dr. Thomas E. Murphy, Photonics Research Laboratory, University of Maryland, USA)

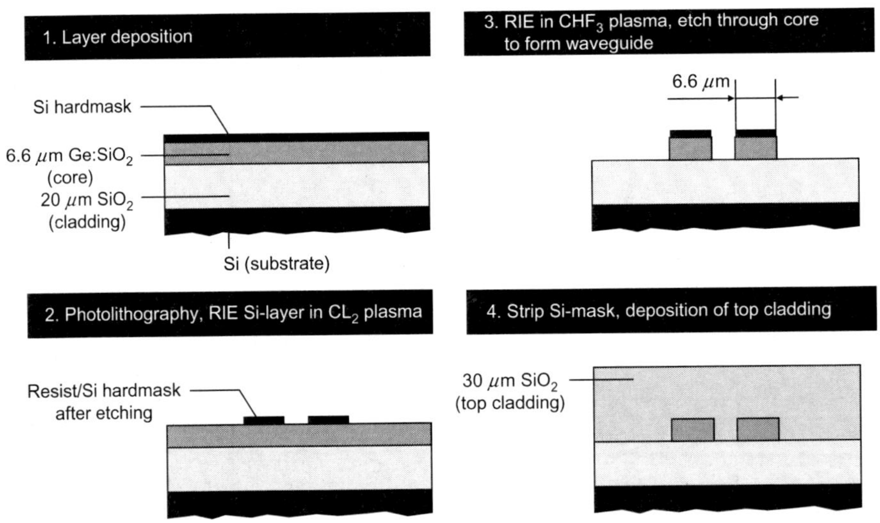

FIGURE 3.12

Processing sequence of advanced waveguide device. (Figure redrawn from reference [56], with permission by Dr. Thomas E. Murphy, Photonics Research Laboratory, University of Maryland, USA)

side walls, but also a sophisticated layout during manufacture of the electron-beam-patterned photomask. Frequently, photomasks suffer from raster steps in the case of shaping the vector scan of the electron beam along the curved segments of the design. This effect results from the way the pattern generator operates; thus the mask fabrication process for curved optical elements has to be optimized for side wall roughness. In general, Gaussian electron beam lithography is preferred for the generation of photomasks containing micro-optical features.

3.8 INTEGRATED OPTICS IN THE MST FOUNDRY SERVICE INDUSTRY: A CASE STUDY

The processes described in Section 3.7 can be supplied by a MST foundry. Examples of services to supply such micro parts can be found on the internet, one of which is described below. LioniX BV is a spin-off company of the MESA+ Institute for Nanotechnology, University of Twente, Enschede, The Netherlands. LioniX provides their customers with product development stage and production services. They are well-known as an optical MST/MEMS foundry, and a leading provider of small to high volume production services of highly innovative products. The company often takes the position of a technology developer, carrying on developments that originated from

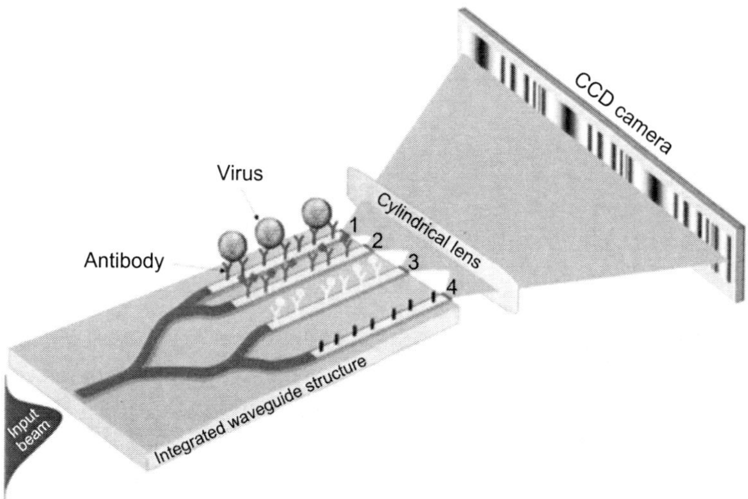

FIGURE 3.13

Illustration of an integrated optical sensor concept. (Reproduced with permission from reference [57]. Copyright ©2007 American Chemical Society.)

research. Bringing such ideas to commercialization requires a distinct skill set. Similar to the concept of a conveyor belt, LioniX can sustain this complex pathway from idea to industry.

Facilitating technology transfer from the academic to the industrial domain plays a large role in the success of high-tech innovation, and LioniX cooperates with their clients and partners through subcontracting, licensing of IP or joint ventures. One example is the integration of optical read-out functionality in biosensors, which was originally developed by the University of Twente [36]. The heart of such a sensor is a Mach–Zehnder interferometer, similar to the design illustrated in Figure 3.11. The portable design of the integrated sensor is its unique selling point.

These sensors may be applied in food or environmental sensing, but higher economical margins may be gained for such an advanced sensor design in the health care market. An outline application is presented in Figure 3.13, and the concept has been published in various scientific journals. The figure is reprinted from *Nano Letters*, Ymeti et al. [57]. More information concerning LioniX foundry services may be found from the company's website [58].

3.9 CONCLUSIONS

Micro- and nanomolding technology for ceramic and organic–inorganic hybrid materials is an emerging field that can be applied to make devices with specific properties including, for example, extreme chemical inertness, bioactivity or high temperature resistance. Many materials are currently being investigated, particularly fused silica and compound semiconductors, which can be used for many optical devices [59–61]. Patterning by RIE is the fabrication method of choice, although many micro- and nanomolding processes are discussed in the literature.

Most of the processes described in this chapter are based on soft molds utilizing PDMS and precursor suspensions. However, techniques such as LIGA and DRIE were highlighted before moving towards more detailed descriptions. In certain applications the use of a hard mold instead of the flexible PDMS may offer certain benefits in production processes, and silicon and replicated polycarbonate have been used. Micromolding in capillaries is discussed, as it may present a particularly promising method for the exploration of functions facilitated by the extensive tailorability of ceramic and organic–inorganic hybrid materials. Micro- and nanomolds can be fabricated by many techniques, including milling, and time- and cost-consuming focused electron and ion beam processes, but generally lithographic processes are needed to produce a high resolution.

One popular lithographic technique is photolithography using SU-8 thick film resist. Using these pre-patterned micromachined silicon wafers allows complex shapes to be realized, such as microneedles. These structures can be used as a master to form a production mold for a subsequent replication process in ceramics.

In all ceramic microprocesses that use precursors, the suspension's composition, drying and sintering conditions will all affect the result of the ceramic molding process. Pressurized filling and long hydration times can improve the results. Ceramic fillers from suspensions suffer from a certain amount of shrinkage, but physical vapor deposition of the ceramic building blocks produces no shrinkage. Self-assembling monolayers were used as anti-sticking layers to allow separation of the hard mold from the ceramic film after deposition. However, the ceramic films may be extremely brittle and may be damaged by demolding, so a backing layer needs to be applied to release the films from the mold. Soft mold replication is easier and allows patterning on curved surfaces due to its flexibility. Soft lithography has already proved to be an outstanding alternative method to conventional techniques. Ceramic materials can be patterned at the nanoscale using this approach, if high temperature treatments are avoided after demolding, since these lead to reflow and shrinkage.

With respect to applications, the tunable porosity of ceramics may offer additional advantages compared to other materials allowing, for example, tissue ingrowth and controlled release of drugs.

Surface micromachining has influenced the rapid rise of optical micro-optical-electromechanical systems (MOEMS). Planar processed features, like mirrors, shutters or Fresnel lenses can be assembled into 3D devices to generate novel MOEMS [27, 62]. The fabrication of optical devices, including waveguides, relies on the same principles as polysilicon surface technology. To minimize optical loss (from attenuation and scattering) in channel waveguides, the fabrication method is closely controlled for good side wall definition of the resist patterning process, and low surface and side wall roughness. Certain optical slab materials may confer a higher etch selectivity using hard masks instead of a polymer (resist) layer. Because these hard masks must be patterned, too, the number of process steps is increased, which could lead to a higher failure rate across the entire process sequence.

In conclusion, we have introduced the processing of speciality substrates, and the concepts involved in the fabrication of integrated lightwave circuits and sensors. We have also briefly discussed the industrial scope of optical integrated microsystems and their applications. In this context, an example of a successful MST foundry service – an illustration of effective technology transfer from universities to the market place – has been presented.

References

[1] G. Feiertag, W. Ehrfeld, H. Freimuth, H. Kolle, H. Lehr, M. Schmidt, et al., Fabrication of photonic crystals by deep x-ray lithography, *Appl. Phys. Lett.* **71** (11) (1997) 1441–1443.

[2] LIGA – Our Key Technology, Helmholtz Zeutrum Berlin, http://www.helmholtz-berlin.de/forschung/grossgeraete/azm/hzb-intern/verfahren_en.html. (accessed 04.04.11).

[3] R.W. Boswell, A.J. Perry, M. Emami, Multiple confined diffusion plasma produced by 13.56 MHz electrodeless source, *J. Vac. Sci. Technol. A* **7** (6) (1989) 3345–3350.

[4] S. Pang, Application of dry etching to microsensors, field emitters, and optical devices, in: P. Rai-Choudhury (Ed.), *Handbook of Microlithography, Micromachining, and Microfabrication,* vol. 2, Ch. 3, pp. 99–152, SPIE Optical Engineering Press, Monograph PM40 Washington, ISBN 0-8194-2379-3(v.2), 1997.

[5] J.H. Keller, New and possible future trends in inductively coupled plasmas (ICPs) for semiconductor processing, *Plasma Phys. Control. Fusion*, Spec. Iss. **39** (5A) (1997) 437–443.

[6] S.A. McAuley, H. Ashraf, L. Atabo, A. Chambers, S. Hall, J. Hopkins, et al., Silicon micromachining using a high-density plasma source. *J. Phys. D-Appl. Phys.* **34** (18) (2001) 2769–2774.

[7] A.M. Hynes, H. Ashraf, J.K. Bhardwaj, J. Hopkins, I. Johnston, J.N. Shepherd, Recent advances in silicon etching for MEMS using the ASETM process, *Sens. Actuators* **74** (1999) 13–17.

[8] M.J. de Boer, J.G.E. Gardeniers, H.V. Jansen, E. Smulders, M.-J. Gilde, G. Roelofs, et al., Guidelines for etching silicon MEMS structures using fluorine high-density plasmas at cryogenic temperatures, *J. Microelectromech. Syst.* **11** (4) (2002) 385.

[9] M.A. Auger, P.L. Schilardi, I. Caretti, O. Sanchez, G. Benitez, J.M. Albella, et al., Molding and replication of ceramic surfaces with nanoscale resolution. *Small* **1** (2005) 300–309.

[10] S.L. Colston, P. Barnes, H. Freimuth, M. Lacher, W. Ehrfeld, Cements: A new medium for micro-engineering structures? *J. Mater. Sci. Lett.* **15** (1996) 1660–1663.

[11] S.L. Colston, D. O'Connor, P. Barnes, E.L. Mayes, S. Mann, H. Freimuth, et al., Functional micro-concrete: The incorporation of zeolites and inorganic nano-particles into cement micro-structures. *J. Mater. Sci. Lett.* **19** (12) (2000) 1085–1088.

[12] E. Kim, Y. Xia, G.M. Whitesides, Micromolding in capillaries: Applications in material sciences, *J. Am. Ceram. Soc.* **118** (1996) 5722–5731.

[13] S.-J. Ahn, J.H. Min, J. Kim, J. Moon, Process mechanism for vacuum-assisted microfluidics lithography with ceramic colloidal suspensions, *J. Am. Ceram. Soc.* **91** (2008) 2143–2149.

[14] O. Guillon, C. Jamin, R.K. Bordia, Effect of drying conditions on patterned ceramic films processed by soft micromolding, *J. Ceram. Soc. Jpn.* **118** (2010) 321–325.

[15] D.-H. Lee, K.-H. Park, L.-Y. Hong, D.-P. Kim, SiCN ceramic patterns fabricated by soft lithography techniques, *Sens. Actuators A: Phys.* **135** (2007) 895–901.

[16] H. Nagata, S.W. Ko, E. Hong, C.A. Randall, S. Trolier-McKinstry, Microcontact printed $BaTiO_3$ and $LaNiO_3$ thin films for capacitors, *J. Am. Ceram. Soc.* **89** (2006) 2816–2821.

[17] Z. Zhu, X. Wei, K. Jiang, A net-shape fabrication process of alumina micro-components using a soft lithography technique, *J. Micromech. Micro Eng.* **17** (2007) 193–198.

[18] Z. Zhu, H. Hassanin, K. Jiang, A soft moulding process for manufacture of net-shape ceramic microcomponents, *Int. J. Adv. Manuf. Technol.* **47** (2010) 147–152.

[19] A.H. Cannon, A.C. Allen, S. Graham, W.P. King, Molding ceramic microstructures on flat and curved surfaces with and without embedded carbon nanotubes, *J. Micromech. Micro Eng.* **16** (2006) 2554–2563.

[20] A.H. Cannon, W.P. King, Microstructured metal molds fabricated via investment casting, *J. Micromech. Micro Eng.* **20** (2010) 025025 (7pp).

[21] J.E. ten Elshof, O.F. Göbel, S.U. Kahn, Micrometer and nanometer-scale parallel patterning of ceramic and organic–inorganic hybrid materials, *J. Eur. Ceram. Soc.* **30** (2010) 1555–1577.

[22] T.H. Lee, Q.Y. Tong, Y.L. Chao, L.J. Huang, U. Gösele, Silicon on quartz by a smart cut process, in: S. Chrisstoloveanu (Ed.), *Proceedings of the 8th International Symposium on Silicon-on-Insulator Technology and Devices*, The Electrochemcial Society, Pennington, NJ, 1997, pp. 27–32.

[23] Y. Takagaki, E. Wiebicke, H. Kostial, K.H. Ploog, Fabrication of GHz-range surface-acoustic-wave transducers on $LiNbO_3$ using imprint lithography, *Nanotechnology* **13** (2002) 15–17.

[24] ATIP document id: 010821ar attip01.038: MOEMS for telecommunications in europe. http://www.atip.org/atip-publications/atip-reports/2001/7354-atip01-038–moems-for-telecommunication-in-europe.html, 2001 (accessed 19.06.11).

[25] R.R.A. Syms, Surface tension powered self-assembly of 3-D micro-optomechanical structures, *J. Microelectromech. Syst.* **8** (1999) 448–455.

[26] L.Y. Lin, J.L. Shen, S.S. Lee, G.D. Su, M.C. Wu, Microactuated micro-XYZ stages for free-space micro-optical bench, in: *MEMS97, Proceedings, IEEE, 10th Annual International Workshop on Micro Electro Mechanical Systems*, NJ, USA, 1997, pp. 43–48.

[27] M.C. Wu, L.-Y. Lin, K.S.J. Pister, S.-S. Lee, Micromachined free-space integrated micro-optics, *Sens. Actuators A: Phys.* **50** (1–2) (1995) 127–134.

[28] C.R. Giles, D.J. Bishop, V. Aksyuk, MEMS for light-wave networks, *MRS Bull.* **26** (2001) 328–329.

[29] R. Sawada, E. Higurashi, T. Ito, Highly accurate and quick bonding of a laser-diode chip onto a planar lightwave circuit, *Prec. Eng. J. Int. Soc. Prec. Eng. Nanotechnol.* **25** (2001) 293–300.

[30] R.G. Hunsperger, *Integrated Optics*, Springer, Berlin, ISBN 3-540-59481-7, 1995.

[31] D. Gloge, Optical waveguide transmission, *Proc. IEEE58* (1970) 1513–1522.

[32] J. Ihlemann, D. Schäfer, Fabrication of diffractive phase elements for the UV-range by laser ablation patterning of dielectric layers, *Appl. Surf. Sci.* **197–198** (2002) 856–861.

[33] R. Krishnan, A. Lisfi, M. Guyot, V. Cagan, Preparation and some properties of pulsed laser deposited $YFeO_3$ films, *J. Magn. Magn. Mater.* **147** (3) (1995) 221–224.

[34] D.A. Nolan, V.A. Bhagavatula, C. Lerminiaux, Integrated-optics planar components, *IEEE Commun. Mag.* **32** (1994) 62–67.

[35] K. Wörhof, A. Driessen, P.V. Lambeck, L.T.H. Hilderink, P.W.C. Linders, Th.J.A. Popma, Plasma enhanced chemical vapor deposition silicon oxynitride optimized for application in integrated optics, *Sens. Actuators A* **74** (1–3) (1999) 9–12.

[36] R.G. Heideman, P.V. Lambeck, Remote opto-chemical sensing with extreme sensitivity: Design, fabrication and performance of a pigtailed integrated optical phase-modulated Mach–Zehnder interferometer system, *Sens. Actuators B* **61** (1999) 100127.

[37] H. Nishihara, M. Haruna, T. Suhara, *Optical Integrated Circuits*, McGraw Hill, New York, ISBN 0-07-046092-2, 2000.

[38] T. Tamir, (Ed.), *Integrated Optics*, Springer, Berlin, ISBN 0-3870-9673-6, 1979.

[39] L. Martinu, D. Poitras, Plasma deposition of optical films and coating: A review, *J. Vac. Sci. Technol. A* **18** (6) (2000) 2619–2649.

[40] D.R. Beltrami, D. Love, A. Durandet, A. Samo, C.J. Cogswell, Fabrication and characterization of a planar gradient-index, plasma-enhanced chemical vapor deposition lens, *Appl. Opt.* **36** (28) (1997) 7143–7149.

[41] T.W. Kim, Y.S. Yoon, Microstructural properties of ZnO epitaxial films grown on p-InP (100) substrates at low temperature, *J. Cryst. Growth* **212** (2000) 411–415.

[42] D. Bosc, N. Devoldre, M. Bonnel, J.L. Favennec, D. Pavy, Hybrid silicapolymer structure for integrated optical waveguides with new potentialities, *Mater. Sci. Eng. B* **57** (1999) 155–160.

[43] C. Darraud-Taupiac, V. Binsangou, J.L. Decossas, J.C. Vareille, Optical waveguides fabricated on polymer substrates by electron beam, *Mater. Sci. Semicond. Process.* **3** (5) (2000) 363–365.

[44] J.H. Kim, E.J. Kim, H.C. Choi, C.W. Kim, J.H. Cho, Y.W. Lee, et al., Evaluation of fluorinated polyimide etching processes for optical waveguide fabrication, *Thin Solid Films* **341** (1999) 192–195.

[45] M. Kawachi, Recent progress in silica-based planar lightwave circuits on silicon, *IEEE Proc. Optoelectron.* **143** (5) (1996) 257–262.

[46] J.L. Vossen, W. Kern, *Thin Film Processes II*, Academic Press, New York, ISBN 0-12-728251-3, 1991.

[47] N. Kaliwoh, J.Y. Zhang, I.W. Boyd, Titanium dioxide films prepared by photo-induced sol-gel processing using 172 nm excimer lamps, *Surf. Coat. Technol.* **125** (1–3) (2000) 424–427.

[48] S.P. Lansley, Diamond photodetctor for deep ultra-violet applications, PhD thesis, University College, University of London, UK, 2001.

[49] S.-D. Cheng, Y. Zhou, C.H. Kam, Y.L. Lam, Y.C. Chan, W.X. Que, et al., Sol-gel derived thin films of $LiTaO_3$ on SiO_2/Si substrates for optical waveguide applications, *Fiber Integr. Opt.* **20** (2001) 45–52.

[50] I. Huttel, J. Gurovic, F. Cerny, J. Pospisil, Carbon and carbon nitride planar waveguides on silicon substrates, *Diam. Relat. Mater.* **8** (2) (1999) 628–630.

[51] M.K. Smit, Integrated optics in silicon-based aluminum oxide, PhD thesis, Delft University of Technology, Delft, The Netherlands, 1991.

[52] J. Zhang, Z. Ren, R. Liang, Y. Sui, W. Liu, Planar optical waveguide thin films grown by microwave ECR PECVD, *Surf. Coat. Technol.* **131** (1) (2000) 116–120.

[53] R.M. de Ridder, R.A. Wijbrans, H. Albers, J.S. Aukema, P.V. Lambeck, H.J.W.M. Hoekstra, et al., A spot-size transformer for fiber-chip coupling in sensor applications at 633 nm in silicon oxynitride, *LEOS95 Annual Meeting*, San Francisco, 1995.

[54] A. Nathan, Y. Bhatnagar, A. Vadekar, W. Huang, Fabrication of a silicon Mach–Zehnder interferometer for mechanical measurands, in: *Proceedings of SPIE - The International Society for Optical Engineering*, vol. **1793**, University of Waterloo, Waterloo, Ontario, Canada, 1993, pp. 19–26.

[55] K. Benaissa, A. Nathan, Silicon anti-resonant reflecting optical waveguides for sensor applications, *Sens. Actuators A: Phys.* **65** (1) (1998) 33–44.

[56] T.E. Murphy, Design, Fabrication and Measurement of Integrated Bragg Grating Optical Filters, PhD thesis, Massachusetts Institute of Technology, 2001, © Massachusetts Institute of Technology, 2001, Chapter 3, pp. 108–120.

[57] A. Ymeti, J. Greve, P.V. Lambeck, T. Wink, S.W.F.M. Van Novell, T.A.M. Beumer, et al., Fast, ultrasensitive virus detection using a young interferometer sensor, *Nano Lett.* **7** (2) (2007) 394–397.

[58] Home page of LioniX. www.lionixbv.com, (accessed 04.04.11).

[59] R.J. Shul, G.B. McClellan, R.D. Briggs, D.J. Rieger, S.J. Pearton, C.R. Abernathy, et al., High-density plasma etching of compound semiconductors, *J. Vac. Sci. Technol. A – Vac. Surf. Films* **15** (3) (1997) 633–637.

[60] R.J. Shul, G.A. Vawter, C.G. Willison, M.M. Bridges, J.W. Lee, S.J. Pearton, et al., Comparison of plasma etch techniques for III-V nitrides, *Solid-State Electron.* **42** (12) (1998) 2259–2267.

[61] M. Schaepkens, G.S. Oehrlein, A review of SiO_2 etching studies in inductively coupled fluorocarbon plasmas, *J. Electrochem. Soc.* **148** (3) (2001) C211–C221.

[62] N.C. Tien, Micro-optical inertial sensors using silicon MEMS, in: Proceedings of the Aerospace Conference, *IEEE 1998*, Snowmass at Aspen, Co, USA, March 21–28, 1998, vol. 1, IEEE, Piscataway, NJ, USA, 1998, pp. 437–443.

CHAPTER

Nanotechnology

4

CHAPTER CONTENTS

Microfabrication for Industrial Applications. DOI: 10.1016/B978-0-8155-1582-1.00004-6

At the same time as processes to shape matter at ever smaller scales, starting with the work piece (Chapters 2 and 3) were developing, another trend appeared. The first set of techniques and processes are referred to collectively as the *top-down* approach; the new trend considers manufacturing from the *bottom-up*, relying on processes known from materials sciences and chemical synthesis. This is not new – nature has become a strong source of inspiration for designers who follow the bottom-up approach in many fields. Process development in the engineering sciences has not only focused on copying from the large to the small, but has also borrowed design strategies from nature and made them into novel, artificially created functions. The latter is often called biomimetics or biomimickry, and has found its way into a range of applications, from self-healing materials in aircraft design to reconstructive medicine. Eventually, the combination of top-down and bottom-up techniques will allow for even more functionality to be combined within a single miniaturized device.

The basic principles of such advanced fabrication methods are the processes of self-assembly and replication. In this context, replication is often referred to as templating. Today, artificially produced devices that can interact intimately with natural body functions are regularly used to replace lost abilities, and have become fully accepted in society. Some examples are artificial limbs, artificial heart valves and vascular-type systems, stents, heart pace makers, cochlear implants, as well as other electrodes which interact with the brain or nervous system such as retina implants and muscle stimulators. At the interface between articifical and biological materials (e.g., implant surface) bionanotechnology gains a particular importance. As well as biomedical engineers, specialists in other fields benefit from these nanoscale developments, for example, in the microelectronics industry (polymer electronics, single-molecule transistor, etc.), or in the further development of novel, advanced microfabrication processes.

In general, better more cost-effective fabrication technology is one of the important drivers for innovation that will lead to an increasing number of novel industrial applications. It is important to stress that nanotechnology is not a single discipline but a field of technology facilitating either the controlled creation, the study of phenomena, or the manipulation of matter at nanoscale dimensions. The fact that highly ordered material systems can be created within a film or bulk structure at the nanoscale facilities often a novel working mechanism but it is the order maintained across multiple length scales that is directly related to control of a system's functions. One can conclude that it is this advanced, man-made, control of functionality down to the nanoscale which characterizes the field of nanotechnology across all scientific disciplines.

A general overview of nanofabrication is given in the book *Micro-Nanofabrication* by Zheng Cui, which presents the principles, capabilities and limits of nanofabrication in the current context of microsystems technologies [1]. I have also written recent reviews on nanolithography for parallel fabrication of nanoarrays in general, and oxide nanoarrays in particular [2, 3]. The following sections will provide an introduction into the field of nanotechnology with specific reference to ordered systems (arrays) and then an overview for further reading on the topic.

4.1 TOP-DOWN, BOTTOM-UP

Initially, there was only one philosophy: the philosophy of miniaturization. In engineering terms, this means that the individual building blocks of a system will become smaller and smaller in size, or will replaced by integrated components requiring less space, since assembly is now omitted. The layers deposited throughout a variety of processes are now definable atom-by-atom. This is a tremendous paradigm shift in microtechnology. Consequently, adding the entire field of materials engineering to microsystems technology demands a clarification of the terminology that is used in this field. Historically, electrical and mechanical engineering relied on *top-down* techniques, while materials and chemical engineering worked from the *bottom-up* and

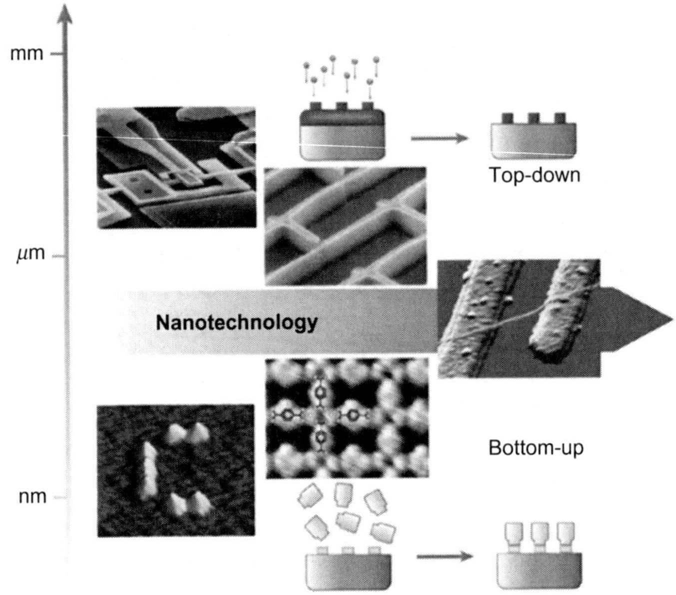

FIGURE 4.1

Top-down and bottom-up nanotechnology. In nanolithographic methods, generating structures at the nanoscale from the top-down maskless or through a mask, patterning can be achieved by writing or printing (from a mask or mold). When a mold is used for imprinting the technique is sometimes also called stamping. In nanotechnology from the bottom-up by self-organization of species, like molecules and solid-state architectures (nanocrystals, quantum dots) from the atomic to the mesoscopic scale, we can distinguish either pure self-assembly or combination thereof with a predefined template generated from the top-down. If such a combination is used we often also call these techniques nanolithography. Shown (clockwise from top) are: an electron microscopy image of a nanomechanical electrometer obtained by e-beam (top-down); patterned film of carbon nanotubes obtained by microcontact printing and catalytic growth (template-assisted bottom-up); a single carbon nanotube connecting two electrodes (electrodes by top-down, whereas positional assembly of the carbon nanotube onto the electrodes is performed manually utilizing the tip of a scanning probe); a regular metal–organic nanoporous network integration of iron atoms and functional molecules, where the iron atoms act as a template in the coordination of the molecules (self-organization from the bottom-up, sometimes also called coordination chemistry); and seven carbon monoxide molecules form the letter 'C' positioned with the tip of a scanning tunneling microscope. (Image originally taken from http://www.physics.ubc.ca/stm/, reprinted from [4] by permission from Macmillan Publishers Ltd: Nature ©2002.)

aimed at modifying a material's properties. In real-world systems, experts from both fields combine their separate approaches, and *nanotechnology* could be considered as a meeting point, where the focus is on the definition of features and processes (e.g., electron transfer) at a scale of a few nanometers. Figure 4.1 illustrates these two opposite approaches to nanotechnology, top-down and bottom-up, with examples of their applications.

4.1.1 Nanolithography

This section provides an overview of nanolithographic techniques, focusing on nanoarrays. Using patterning techniques such as lithography, we normally aim for a higher order architecture similar to functional systems in *nature*. Inspired by its wealth of complexity these architectures are translated into technical devices such as, for example, those found in integrated circuitry or other systems in which structural elements work as discrete building blocks in microdevices. Ordered artificial nanostructures (arrays of pillars, holes and wires) have shown particular properties, and bring about the opportunity to modify and tune the device operation. Moreover, these nanostructures open new applications, for example, the nanoscale control of spin direction within a nanomagnet. We can then look for applications where this unique property of the smallest manufactured element is repetitively used – with respect to spin, this would be in nanopatterned magnetic media for data storage.

These highly ordered, parallel manufactured nanostructures are generally called *nanoarrays*. Most of these applications require massively parallel produced nanopatterns which can be directly realized by laser interference (areas up to 4 cm^2 are easily achieved with a Lloyd's mirror set-up). In this overview we will further highlight the application of laser interference as a tool for nanofabrication, and its utilization as a primary patterning method for nanoimprint applications. Laser interference has great potential for producing simple (dots, lines, ellipses) and periodically repetitive patterns which could potentially be used in a range of devices, including nanostructuring for photonic crystal devices, high resolution patterned media, and surface modifications of medical implants. Laser interference lithography on the other hand clearly has limitations in terms of pattern freedom.

The unique properties of nanostructured surfaces have found applications in biomedical nanoarrays, either for diagnostic or functional assays, including catalytic reactions on chip. Bio-inspired, templated nanoarrays will be discussed in terms of how they fit with other massively parallel nanolithographic techniques currently under development in nanotechnological applications.

4.1.2 Introduction to the Need for New Lithographic Techniques

The vast amount of publications concerning lithographic techniques, which are capable of forming structures with lateral dimensions below 100 nm, suggest that high-end exposure techniques are well established. Particularly, in microelectronic systems, production machines are currently under development which are capable of reaching the 32 nm node, which is defined as the average half-pitch within a semiconductor pattern. Ultimate resolution, i.e., controlling functional elements at the molecule, atom, electron or spin scale, remains the major goal for further lithography research. This research field has consequently been branded *nanolithography*. A selection of nanolithographic techniques are of particular interest for their ability to produce massively parallel, repetitive nanostructures. Each technique needs to be evaluated on the basis of the following three criteria:

1. capability of the lithographic technique to resolve lateral features at a scale below 500 nm,
2. provision of an extremely high ordering across the desired manufacturing area, and
3. distinct positional control of placing, and ability to address an individual feature within the system.

A single feature reduced to the nanoscale (e.g., a dot) may by itself already possess remarkable properties, like unique electron transport properties, that are yet to be fully discovered [5]. Applications which make use of the repetitive nature of uniform nanostructures across a large manufacturing area have already demonstrated their potential, e.g., in patterned media (PM) for data storage [6].

This overview is a compilation of the literature which discussed major recent nanolithographic principles used in the pattern transfer of these massively parallel nanostructures. It is important to distinguish these nanoarrays with geometrical features at the nanoscale from biological screening arrays containing nanolitre-sized wells, which are also called nanoarrays [7].

Often the research on nanolithography for the fabrication of a nanoarray is dedicated to a specific application field, such as the fabrication of data storage media, photonic crystals and nanoarrays for diagnostics. For a better understanding of how these lithographic techniques may be characterized, Figure 4.2 shows the relation between the length scales of interest and their extreme demands on placement accuracy to achieve the required high ordering of the nanofeatures. Ideally, nanoarrays can be produced across the

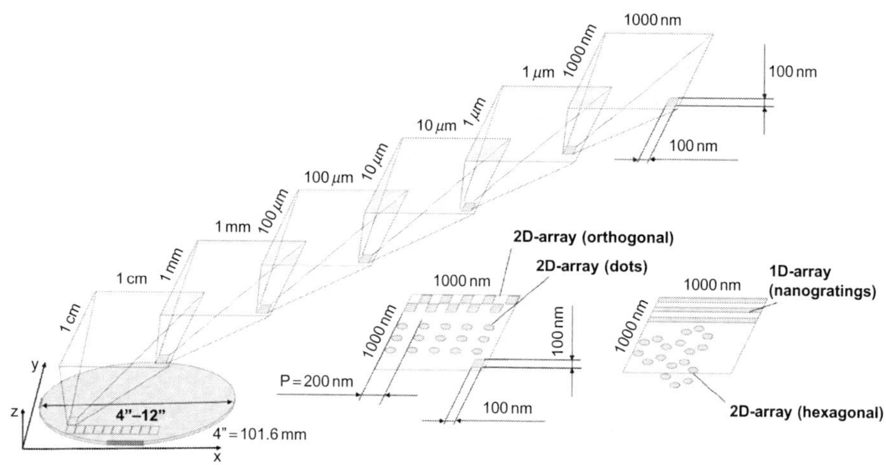

FIGURE 4.2

Definition of lithographically manufactured nanoarrays consisting nanofeatures at the scale of 100 nm. (Reproduced from reference [2] with permission from IOP Publishing.)

full area of an industrial scale wafer surface – generally a diameter of 6 inches or more, although for many devices such extreme large-area and continuous patterning may not be required. Considering that defect-free patterning across the area of a typical microelectronic silicon die of 26×33 mm^2 is already a major challenge, we define here an area of 1 cm^2 as large.

Figure 4.3 presents two kinds of nanoarray layouts: the continuous and the discontinuous nanoarray, where the latter is defined by repetitive clusters of nanoscale features that in themselves are also ordered. Comparing nanoarrays to a continuous thin-film layer of the same material can deliver novel properties to the layer. One example of this is a photonic crystal. Such crystals have tunable optical properties for guiding light within them, in a way that is defined by the lateral feature size, shape and periodicity of the individual nanostructures. These nanostructured materials are often called metamaterials (see section 4.2).

Finally, we need to define the difference between the lithography and the pattern transfer step (see also Chapter 2). Generally, transfer means that a pattern is carried over from one layer to another. These layers are either chosen for their masking, functional, or sacrificial nature, such as for example a silicon nitride layer on silicon, where the nitride layer protects the silicon during etching.

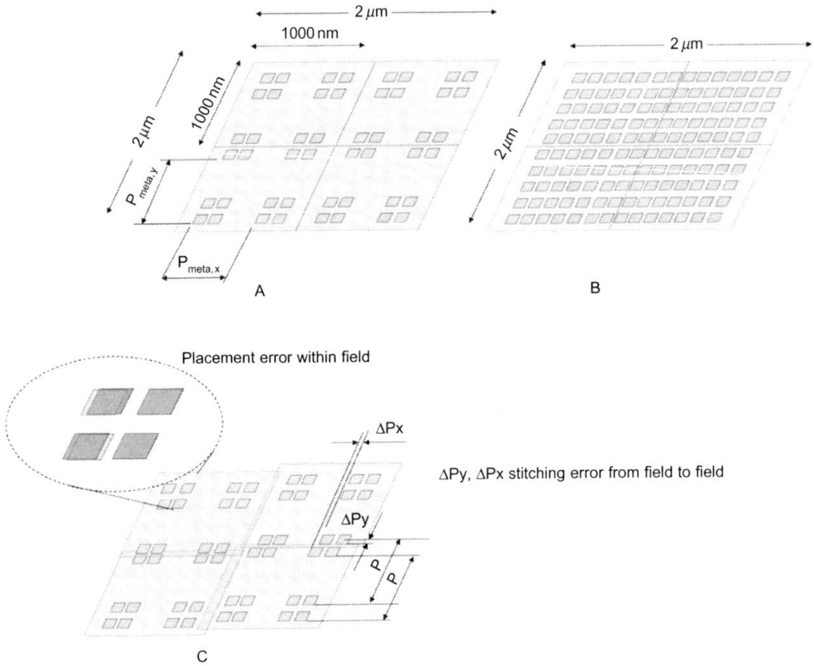

FIGURE 4.3

Definition of (A) discontinuous and (B) continuous nanoarrays, and (C) possible errors. Both (A) and (B) can deliver metamaterial properties dependent on the combination of the dimension of the single feature and the meta-period (Pmeta) of its repetition within the array. A field is defined here by $1000 \times 1000\,nm^2$. Alignment errors, positioning from field to field (stitching), overlay errors and placement errors within one field can occur. These errors will depend on the lithographic technique and are not generally defined. A typical microelectronic chip (die) covers, e.g. an area of $26 \times 33\,mm^2$. In the semiconductor industry a roadmap is assembled to illustrate lithographic needs for innovation. A novel lithography technique must guarantee a certain degree of fault-free structuring within the patterning fields across the die. Of course, the exact definition of feature tolerances highly depends on the function of a device. For semiconductor devices a summary of these definitions can be found at [10]. (Reproduced from reference [2] with permission from IOP Publishing.)

Typically, lithography employs a resist layer. This layer is normally a polymer that is sensitive to light of a certain wavelength and etch-resistant in the transfer step – hence the name resist. Recently, a range of inorganic resists have been indentified, for example, sol-gel based resists in nanoimprint lithography [8]. The pattern is transferred from the mask to the resist by an exposure step, from a resist to an underlying functional layer by etching, or from a resist to an added layer by deposition and lift-off. Many varieties of process sequence for pattern transfer are known and several extensive texts are available, for example, *Handbook of Microlithography, Micromachining and Microfabrication* [9].

At a length scale below the lateral width of 500 nm, it becomes extremely difficult to fabricate structures from the top-down using 365 nm-ultraviolet (UV) photolithographic techniques through a mask, because of diffraction of the light at the mask openings. Moving to a shorter wavelength, specifically deep ultraviolet (DUV), is practicable, for the sub-500 nm domain, but few such industrial systems are accessible to researchers in the field of nanotechnology. Since these machines are dedicated to electronics, they will not be reviewed in detail here, but briefly discussed in the context of next generation nanolithography principles (section 4.1.3). Features can be made quite reliably, even down to a few nanometers, by electron-beam lithography, however. Current, state-of-the-art electron beam (e-beam) lithography is an indispensable technique for the fabrication of templates (also named masters or molds) for nanotechnological applications. This top-down technique is extremely slow and, as shown in Figure 4.3, this type of pattern delineation produces stitching errors and discontinuities in nanoarrays. Electron beam lithography (EBL) has been considered as an option. It has so far failed to become the flagship of nanotechnological production techniques because it uses a serial write process. On the other hand there are emerging very promising new parallel particle beam lithography systems, using so-called multiple electron-beam/ion-beam sources on the market for maskless lithography, e.g., Mapper Lithography BV [11]. Mapper's machine uses over 10,000 beams to scan the wafer surface for direct write circuit patterns, and aims at 22 nm lithography without the need for expensive masks.

In the field of medical applications a multitude of nanoarrays can be found in use for diagnostic purposes. Functional arrays, for synthesis also exists, for example those which exploit enzyme reactions on a chip. In section 4.1.6 we will discuss nanoimprinting as it is the most popular nanotechnology currently explored in research. Nanoimprint lithography is a top-down, template-assisted, nanolithographic method, which already reaches production standards for cost-effective pattern transfer, exploiting the existing semiconductor mask standards for template fabrication. An overview of self-assembly techniques from the bottom-up, exploring biological substrates such as DNA, is given in section 4.1.8.

If interference is utilized, the initial lithographic step can be performed without a mask or template, which can be an advantage for some applications, e.g. applications with relatively low production volumes. Although only briefly mentioned in section 4.1.9, the development of maskless techniques such as interference lithography (IL), are considered as important alternatives to photo-, e-beam, self-assembly and nanoimprint lithographic techniques, specifically for patterning very large-area periodic nanostructures.

Over the last two decades, many new resist systems have appeared in lithography, and the stringent demands for innovation in the semiconductor industry have fostered tremendous technical improvements in exposure tools. A new exposure tool initially requires the definition of a process line, specifically with respect to nanolithography. An example of this is IL nanofabrication, used at the MESA+ Nanolab NL. Periodic nanostructures are achieved in a single, maskless lithography step and e.g., recording experiments in a diversity of specifically designed PM materials including the development of read/write technology at the nanoscale can readily take place. Beyond this specific application, the MESA+ Laser IL (LIL) system has also been used for a variety of experiments which study photonic crystals, hierarchical nanostructures produced by self-organized (templated) assembly, and also the creation of nanostructured surfaces for the investigation of a novel class of biomaterials exposing nanotopographically designed surfaces to biological systems. These approaches will be discussed in section 4.1.11.

In sum, this overview presents the technological aspects of a selection of nanolithographic processes for nanoarrays. It may also offer a starting point for the further development of specific design rules for massively parallel nanoapplications. This section also discusses the broad field of available nanolithographic routes for the fabrication of nanostructures with a focus on arrays, but is by no means limited to this specifically challenging nanopattern. Choice of nanolithographic fabrication technology depends on the size and shape of the individual nanostructure, and the requirements of the extent of periodicity within the nanoarray; the field of application also has some influence. Various examples of nanoarray applications in information storage and medicine are presented in the following sections.

4.1.3 Nanolithographic Techniques

The basic principles of lithography have already been described in Chapter 2. Figure 4.4 illustrates the four lithographic methods, which use a masking layer, patterned through either lithographic exposure (illumination, A and B),

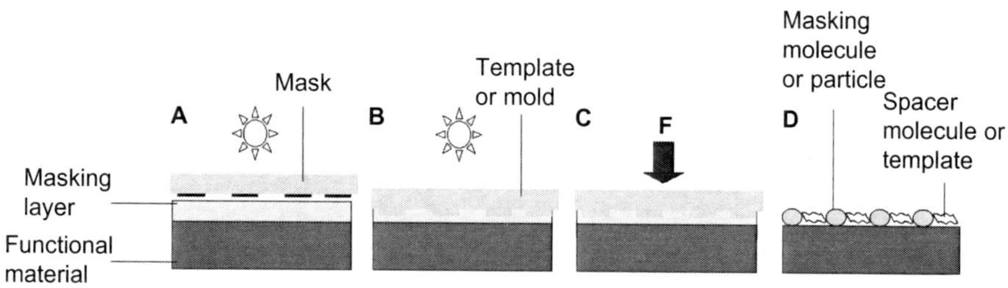

FIGURE 4.4

Principles of lithographic methods. Lithography utilizes in its strict definition a masking layer. A variety of technologies exist to generate such a masking layer. In most cases the mask or template fabrication process by itself is also a lithographic process using a direct-write method, mostly e-beam lithography. Although stencil masks, e.g., fabricated by techniques such as focused ion-milling are possible, they are not broadly established in manufacturing. The most established technique is (A) photolithography through a so-called chromium mask. Due to the resolution limitations in photolithography, often nanofeatures at the nanometer scale are transferred by nanoimprinting either utilizing a wet process, hardening a polymeric layer through UV exposure (B), or by the application of a (C) thermal process pressing a template into a thermoplastic layer. Alternatively, a masking layer can be generated by (D) self-assembly, where in some of these self-assembly nanolithographic techniques the spacer or template molecule can be omitted. The common aspect of lithographic techniques is here defined as the fact that the pattern generated in the masking layer is subsequently transferred into a functional material.(Reproduced from reference [2] with permission from IOP Publishing.)

or imprinting (C) via a photomask or template. In comparison Figure 4.4D depicts the patterning of a masking layer by self-assembly. The actual transfer of the functional material by deposition, lift-off, etching or layer-by-layer deposition is strictly speaking not part of the lithographic step [12]. This difference is often not made clear, since it is automatically assumed that lithography is carried out for the subsequent shaping of the work piece by pattern transfer steps.

Within the field of nanotechnology this distinction between lithography and pattern transfer has been weakened in the exploration of novel nanostructuring methods, particularly since techniques such as scanning probe lithography produce a functional nanostructure directly, e.g., by dip-pen lithography. Several outstanding contributions have been made to the field of direct write scanning probe lithography techniques by the Mirkin Research Group at the Northwestern University Evanston, IL, USA. These researchers utilize multiple cantilevers fabricated by microelectromechanical systems' technology to increase throughput. One of their specific foci is controlling architectures on the 1–100 nm length scale, demonstrating dip-pen massively parallel pattern transfer in nanomedicine, examples of which are described elsewhere[13–16]. It is, however, too early to categorize any of these lithographic techniques in terms of their importance for the field of miniaturization. According to Figure 4.4 it is much more important to ask the question if a masking or templating layer is needed. The choice of a specific transfer process may deliver certain benefits in respect of the realization of the functional properties of a nanoarray. Of course, this conclusion may be valid for many other designs, too.

4.1.4 Top-Down Nanolithographic Principles

The e-beam direct-write approach offers the highest pattern fidelity (freedom of shape, accuracy, and precision of features and overlay) of all existing top-down nanolithographic methods, but it is mainly used to produce masks and template structures because it is slow. In combination with nanoimprinting, however, it provides a versatile and cost-effective route for the fabrication of many interesting nanodevices. So far, no better technique exists in template manufacture. Although laser interference lithography (LIL) is a much faster and cheaper method for maskless massively parallel patterning, it cannot compete with e-beam lithography in waferscale pattern uniformity and freedom of pattern geometry. These drawbacks of LIL may be due to the lack of suitable large-area mirrors or other suitable interference optics, but 193 nm-deep ultraviolet (DUV) and extreme ultraviolet (EUV) stepper and scanner techniques are emerging lithographic technologies for the semiconductor industry that already offer very high precision and resolution.

Since these technologies are mainly restricted to very high volume production only a brief overview will be given here. Work on these new light sources was initiated by the contributions of Bokor and Oldham [17, 18] amongst others. Maskless techniques, including multiple electron/ion-beam lithography and maskless DUV/EUV interference lithography may at some point overcome the limitation of low throughput in e-beam lithography, at least for specific periodic structures. The selection of a nanolithographic manufacturing technique will always be a compromise between machine cost, speed (wafer throughput) and pattern fidelity (resolution and process stability). In high volume production, where throughput is the major criterion for choosing a specific lithography technique, template replication processes may succeed classical photolithography, but the initial costs making a 1:1 template can be high.

This is similar to the situation in the 1980s, when x-ray lithography (lithography with x-rays as the illumination source) was considered to be the next generation lithography technique for optical lithographic techniques. The definition of an appropriate cost model for nanodevices based on any one of the nanolithographic strategies goes beyond the scope of this overview.

A totally different template-free approach to nanoarray fabrication is also worth mentioning. Materials can be manipulated by low-dose electron beams. If a thin-film polymer is exposed to the transmission electron microscope, the real-time, direct manipulation of the polymer chain aggregate structure becomes possible in two dimensions. The polymer melts, and the chains move by attraction toward preferred electron concentrations within the beam. The principle may be compared to other direct energetic beam-induced surface tension gradients yielding the formation of patterns in material by laser interference metallurgy [19, 20] but the work is, as yet, highly experimental. New machine architectures, novel resists and etch-through processes for lithographic pattern transfer are also being developed, as recently reviewed by Argitis et al. [21]. *The Proceedings of SPIE – The International Society for Optical Engineering* is an invaluable information resource for details of novel processes, often those which are linked to commercial facilities, e.g., volume 6923 (2008), *Advances in Resist Materials and Processing Technology XXV*.

4.1.5 Nanolithographic Technologies for the Microelectronics Industry

Deep Ultraviolet (DUV) Lithography

DUV is a key enabling technology which uses 193 nm excimer lasers to produce commercial chips at the 90 nm technology node. Validation of the 193 nm immersion scanner at IMEC's 300 mm wafer fab facility in Leuven, Belgium, is a clear milestone toward ensuring the availability of a cost-effective technology at 65 nm and may even be extended to the 45 nm node for microelectronics. Many publications have been presented concerning this technological development, of which only a few are cited here [22–26]. The development consists of the definition of a suitable light source, optics and other hardware aspects of such a lithography system, including mask technology, immersion liquid and defect analysis [27].

Extreme Ultraviolet (EUV) Lithography

EUV is considered to be one of the key enabling technologies, much as DUV. The technique is hoped to fill the gap in the semiconductor roadmap towards the 22 nm node for production. A recent trend to overcome the absolute limit of the photolithographic method, is to use the "More than Moore" concept [28–35], hence only a few players are developing EUV as a next generation lithography tool [36–41], including IMEC in Europe and the Lawrence Berkeley National Laboratory in CA, USA. The technique is approaching already the 32-nm node generation of very-large-scale integrated (VLSI) electronics by optical lithography utilizing a 248 nm-laser source, hence EUV is still one of the top candidates for industry to adopt in production. Although investment costs are very high EUV offers very reliable access to high resolution nanostructures, which will clearly be able to satisfy the market needs for high volume nanoelectronic device production for some time to come.

4.1.6 Nanoimprint Technology

Some very good papers have been published concerning the development and application of nanoimprint lithography as a massively parallel pattern transfer technique suitable for industry (principle illustrated in Figure 4.4B and C). Evolved from hot embossing at the microscale, thermoplastic nanoimprint lithography was first developed by Chou's group [42–44]. Morton et al. have also demonstrated wafer-scale patterning of sub-40 nm diameter and high aspect ratio (>50:1) silicon pillar arrays by nanoimprint and etching [45]. Further developments in this area of research have led to a combination of imprinting and UV exposure, so-called step-and-flash imprinting. This process has been investigated and brought to production standards contributions mainly by Willson (University of Texas, USA) and Resnick (Molecular Imprints, Inc.) who have together published more than 20 papers concerning step-and-flash imprint lithography [46]. Willson's group continues to develop and optimize materials and processes for this technology [47]. These techniques are at the frontier of nanopatterning and include hybrid techniques such as physical-confined colloidal lithography.

4.1.7 Case Studies: Nanoimprint Applications

Patterned Magnetic Media

Before we start describing the specific nanoimprint processes let us recap some background on patterned magnetic media and the impact of nanoscale in this field of research. Researchers at the Hitachi San Jose Research Center, CA, USA, beautifully presented in *Solid State Technology* a cover article concerning the patterning of hard disk drives [48]. Given the fact that this vision was presented around two years ago, and the first mention of perpendicular patterned media substrates for data storage was made about 10 years ago, we can conclude that technological development needs transit time (time-to-market) to progress from the research laboratory to industry. The development of robust fabrication technologies is a significant part of this progression [49, 50].

Figure 4.5 illustrates the concept of patterned media (PM), comparing single domain magnetic islands with classical multiple grain media, which suggests a solution to the thermal stability problem by the use of a predefined magnetic dot per bit [51]. Realizing dots at a size of 25 nm diameter and periodicity of 35 nm, within a cost-effective high-volume production process is still a major challenge in nanofabrication. UV-cure nanoimprint lithography is one of the emerging technologies that show the capability to achieve both the high required throughput and resolution for this class of device.

Moritz et al. have reviewed the use of PM disk fabrication [53]. They demonstrated ultrahigh-density magnetic recording by two ways of patterning: a) direct e-beam lithography and subsequent reactive ion etching, and b) nanoimprinting. The nanoimprint technology delivers polymer arrays with features as small as 30 nm, with a pitch of 60 nm, which are subsequently coated with a Co-Pt multilayer with perpendicular magnetic anisotropy.

Terris and Thomson have reviewed self-assembled PM technology [54], which also requires the development of novel read/write technology. One approach can also be found in MEMS technology based on a cantilever design for atomic force microscopy originally introduced by Rohrer and Binnig [55, 56]. More recent examples of patterned media are given in a review by Terris et al., who also refer to nanoimprinting as one of the emerging possibilities for the cost-effective nanofabrication of PM disks [57], but the authors do not give an opinion as to whether any of these processes will be cost-effective to introduce.

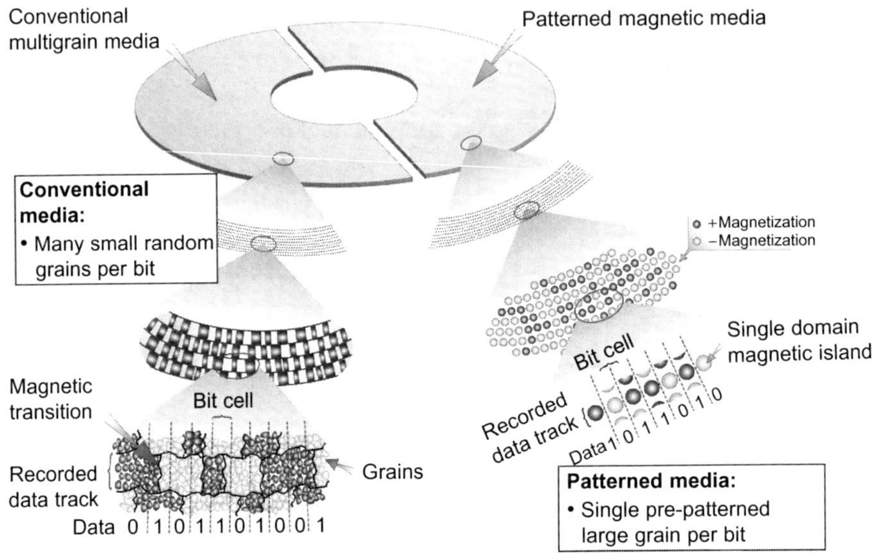

FIGURE 4.5

Overview of granular versus patterned media (PM) for data storage. Comparing granular media with PM, the novel concept consists of an array of magnetic islands created in the form of a dot. One dot represents a single domain, i.e., spin-direction. One dot is either "written" or "non-written". In the creation of spatially separated magnetic islands a higher resolution and lower noise level can be achieved than in conventional media. (Reproduced from reference [52] with permission ©IEEE 2008.)

Random Access Memory

A 6 nm half-pitch lines and 0.04 μm^2 static random access memory (SRAM) pattern by nanoimprint lithography has been demonstrated by Austin et al. [58]. The authors manufactured a specific high resolution mold by cleavage of a GaAs superlattice grown by molecular beam epitaxy. With these type of patterns they reached the limit of reliable SEM inspection, however, it is suggested that the UV curable nanoimprint process will support line width even smaller than 6 nm.

Surface-Acoustic-Wave Devices

Researchers at the Paul-Drude Institut für Festkörperelektronik, Berlin, Germany, have fabricated surface-acoustic-wave (SAW) transducers operating in the GHz range on $LiNbO_3$ using thermal imprint lithography [59]. The width of the metal lines was 170 nm at a period of 600 nm. Although these SAW devices do not challenge the resolution limit of this technology, which is currently of the order of 10 nm, they are an interesting example of devices having large volume applications in telecommunication.

4.1.8 Emerging Nanolithographic Technologies

Multiple E-Beam/Ion-Beam Lithography

This is a rather different approach to the development of DUV and EUV scanners. It follows the idea of upgrading the wafer throughput of direct e-beam/ion-beam lithography by directing multiple beams in parallel across the wafer area [60–68]. These maskless, massively parallel patterning techniques aim for critical dimensions at the 50 nm line width. Several researchers have demonstrated the potential of such a lithography concept by fabricating multiple-beamlet arrays utilizing MEMS technology. It is not yet possible to assess the position of these emerging technologies on the industrial roadmap.

Templated Nanoarrays

Label-free and high-resolution protein/DNA nanoarrays are an important development within medical diagnostics. Sinensky and Belcher described such an array using Kelvin probe force microscopy as a detection technique [69]. With their method of probing surface-bound target biomolecules, they show that nanoarray diagnostics at 10 nm resolution are potentially possible. Although there is no elaboration of nanolithographic fabrication strategies mentioned in their paper, and their choice of dip-pen nanolithography (DPN) over other nanolithography techniques may be arbitrary, the question of what are the specific requirements for this class of nanoarrays remains. Utilization of one of the following templating techniques is quite interesting and may also be suitable for the fabrication of such nanoarrays.

Reaching the resolution limit of physical templating by top-down lithographic techniques many other substrates including biological are explored. Lee et al., for example, reported the ordering of quantum dots using genetically engineered viruses by fabricating highly ordered composite materials in their publication in *Science* [70]. In their paper, highly ordered, A7-ZnS, self-supporting viral films are produced. Figure 4.6 gives an overview of the engineering length scales involved in such film. This example demonstrates the extent of the possiblities for self-assembly. A similar process is presented by Huang et al., who demonstrated a programmable assembly of nanoarchitecture, also using genetically engineered viruses [71].

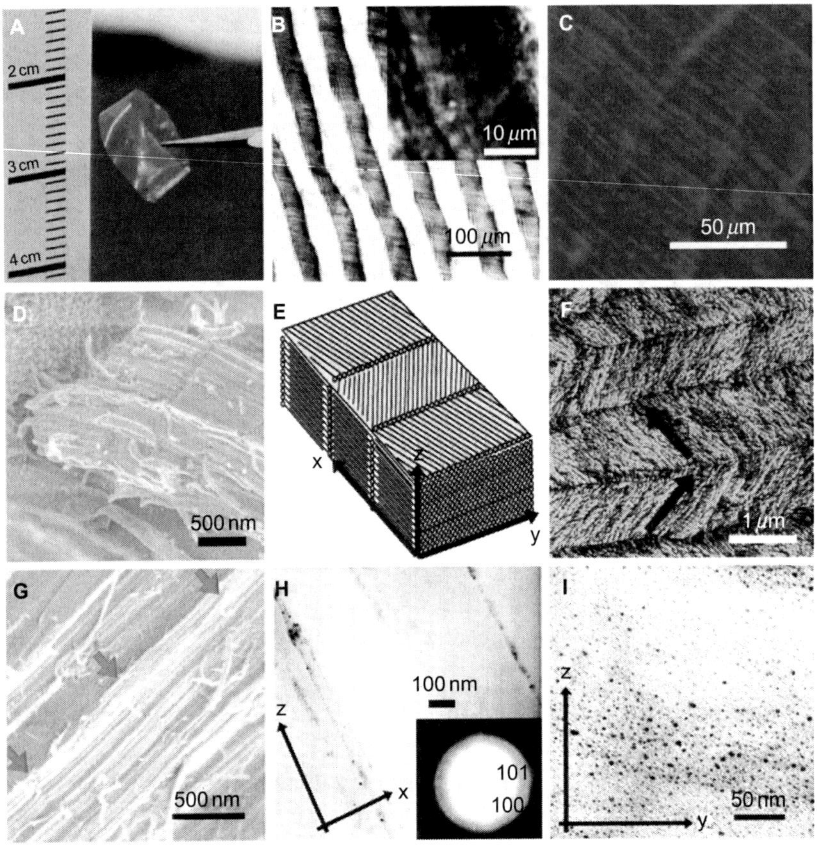

FIGURE 4.6

Characterization of A7-ZnS film. (A) Photograph of A7-ZnS viral film. (B) POM (20) birefringent dark and bright band patterns (periodic length 72.8 μm) were observed. These band patterns are optically active, and their patterns reverse depending on the angles between polarizer and analyzer. (C) Photoluminescent image, with an excitation wavelength of 350 nm and with filtering below 400 nm, shows 1-μm stripe patterns (50). (D) SEM images of highly packed three-dimensional bulk film structure. (E) Schematic structural diagram of the A7-ZnS composite film. (F) AFM image of the free surface. The phage forms parallel aligned herringbone patterns that have almost right angles between the adjacent director (arrows). (G) SEM image showing the close-packed lamellar structure of phage and nanocrystal layers (diagonal, thick arrows) in the inner areas of the film. (H) Low-resolution TEM image of cross section of A7-ZnS film, with 20 nm $\times$ 2 μm ZnS nanocrystal stripe pattern aligned between one phage length in the x–z direction of film; the inset shows an electron diffraction pattern of ZnS wurtzite structure. (I) Low-resolution TEM image of film viewed in the y–z direction, showing ZnS nanocrystals. (Reproduced from reference [70] with permission from the American Association for the Advancement of Science ©2002.)

Alternative templates to phages can be used, such as DNA. In 2006, Feldkamp et al. [193] compiled a review of rational design for DNA nanoarchitectures, in which DNA is used as a building block at the nanometer length scale. This is a technology concept that could offer a large variety of designs for self-assembled supramolecular arrays and other scaffolds, from which it may be possible to form biosynthetic, mechanical and logical nanodevices. To avoid replicating data from this review, we will limit this overview to DNA-based techniques that have been applied to form highly ordered nanoarrays, and have been published over the last few years. For example, Becerril et al. have reported the use of DNA-templates for nickel nanostructures [72]. These nickel nanomaterials may be applied directly as catalytic platforms, or used for the selective binding of protein assemblies. This concept has been demonstrated for the reversible association of a histidine-tagged phosducin-like protein.

Another example of the exploitation of the inherent recognition and self-assembly capabilities of DNA is described by Wang et al. [73]. This group have synthesized and characterized a matrix assembly of single-wall carbon nanotubes (SWNTs), that were functionalized with carboxylic groups and single-stranded DNA (ssDNA) demonstrating negative differential resistance in the biotemplated interfaces of the matrix structure. Combinations of this bio-coordination chemistry will produce a diversity of nanoelectronic effects which have yet to be investigated.

More established fabrication techniques for bio-nanoarrays include the electrochemically controlled assembly of DNA and protein biomolecules onto gold nanoelectrodes, and templating protein assemblies on micro-contact printed surfaces. Huang et al. describe a dynamically configurable biomolecular array, produced by adjusting the electrochemical potential applied at the nanoelectrode allowing for a variety of combinatorial interactions [74]. In other work, Iversen et al. describe a functional nanoarray exploring SNAP-tag proteins affinity bound to patterns of streptavidin modified with BG-biotin that has been stamped onto a surface by micro-contact printing (microCP) [75]. Also electrical-field assisted nanofabrication can be utilized to form multilayered structures composed of biomolecule-derivatized nanoparticles. In the study by Dehlinger et al., a microelectrode-array device facilitates the rapid parallel electrophoretic transport and binding of biotin and streptavidin fluorescent nanoparticles to specific sites on a microarray [76]. Chhabra et al. discuss a spatially addressable multiprotein nanoarray based on aptamer-tagged DNA templated nanoarchitectures [77].

Han Yan's group and their collaborators in the Arizona State University, USA, have published around 50 papers dedicated to DNA-tile-based self-assembly and its application to biosensors. One of these contributions is particularly worth mentioning in the context of nanotechnology, since it describes the phenomenon of multivalency, through which it is possible to assemble multiple molecules while controlling the spacing between them [78]. Another route being explored extensively in self-assembly is the utilization of pore-type structures, of which an overview is given by Hirst et al. [79]. They describe how specific gelator molecules can be assembled for functional nanoscale architectures. Synthetic chemistry that exploits the supramolecular structures that form in molecular gels can be applied as nanostructured scaffold, e.g., for the regrowth of nerve cells, but has also gained interest in nanoelectronics. Use of this type of smart polymeric gel to redefine the limits of biomedical devices has been also described by Chaterji et al., who summarized recent progress in the development and application of polymeric gels based on defining their properties, describing stimuli response and illustrating applications of smartly designed gels to microfluidics [80].

Parallel arrays of geometric nanowells for assembling curtains of DNA with controlled lateral dispersion have been recently described by Visnapuu et al. [81]. In their paper, a process is demonstrated that allows the assembly and visualization of curtains comprising thousands of individual DNA molecules, organized at a diffusion barrier on a lipid bilayer-coated surface. The curtains are made by a combination of geometric barrier pattern with nanofabrication utilizing single-molecule optical microscopy. Unfortunately, the full scale of these fascinating applications of nanoarrays and their fabrication strategies cannot be discussed here. There is, however, much further information available in the literature; for example, Tang et al. describe recognition imaging and highly ordered molecular templating of bacterial S-layer nanoarrays containing affinity tags, and Zhang et al. present the production of nanopatterns by a combination of e-beam lithography and self-assembled monolayers for an antibody nanoarray [82, 83]. In addition, Chen and Li contemplated a record of nanotechnology: moving from microarrays toward nanoarrays [84]. A review of the development of high-throughput methods for functional analysis of diatom genes is given by Montsant et al., and Lynch et al. address functional protein nanoarrays for biomarker profiling [85, 86].

Final mention will be made of the efforts to produce three-dimensional architectures utilizing protein templating. A fairly recent attempt by

Allred et al. describes the three-dimensional architecture of inorganic nanoarrays electrodeposited through a surface-layer protein mask [87].

Geometrically constrained inorganic templates as well as biomolecules have been explored for their potential for templating in nanoarrays. For example, nanofabrication of plasmonic structures is reviewed by Henzie et al. who describe the exceptional optical properties present in arrays generated by soft lithography techniques [88]. Hierarchically structured nickel nanoring and hollow sphere arrays were generated by morphology inheritance based on an ordered through-pore template and electrodeposition by Duan et al. [89]. Dipping an alumina template (AAO) into a colloid solution forms a colloidal monolayer with fine structures on the nanoscale. The formation of such morphologies is attributed to preferential growth along the inner wall of the alumina pores. The fine structure of the nanoflakelet originates from a morphology inheritance process of the transitional product $Ni(OH)_2$, which gives the final nanostructured Ni crystals their desirable magnetic properties. These structures may find application in the fields of catalysis, magnetics, optoelectrics, surface-enhanced Raman scattering (SERS), and other new nanodevices.

The self-organized structure of alumina has been utilized for template formation for some time. Her et al. have demonstrated the fabrication of nanostructured titania [90]. Their paper describes the formation of thin film titania by spin-coating in the sol-gel state and subsequent nanoimprinting using an alumina template to yield a dense array of nanopoles. Nanopores in titania were also achieved by utilizing a suitable mold structure. These nanoarrays may also be useful for solar cell, photocatalytic and sensing applications. Kim et al. have demonstrated that fibroblast cells respond to the nanotopology of polymeric surfaces (PMMA) replicated by AAO nanoimprinting [91]. This area of research is potentially of great interest for tissue regeneration of bone. The investigation of osteoblast response to large-area scalable biomaterial nanotopographies has been facilitated by the lithographic techniques discussed here, specifically laser interference lithography as developed at MESA+ [92].

A large number of combinations of materials have been investigated to determine their potential for producing ordered, self-organized nanodot arrays. For example, Meletis and Jiang described the formation of cobalt nanodots in diamond-like carbon thin films that were deposited by magnetron sputtering in a plasma-assisted Ar/CH_4 discharge [93]. Cobalt dots of 5 nm diameter, separated by 1–2 nm DLC (diamond like carbon) boundaries and arranged in hexagonal arrays, were produced on silicon substrates. They demonstrate ultrahigh resolution. Such well-ordered magnetic nanodots are promising for the next generation of patterned magnetic media. This class of templated nanoarrays also contains nanoscale structures formed by block copolymer deposition, which generally behaves, in its self-organized ordering, in a similar way to submicron inorganic colloidal self-assembly on surfaces [94, 95].

Colloidal self-assembly has also been used in early applications of massively parallel pattern transfer, known as colloidal, particle, block copolymer or nanosphere lithography, depending on the specific materials being used [96, 97]. However, the self-assembly of a thin film of, for example, a spherical-morphology block copolymer is also mentioned as block copolymer micelle nanolithography, and it can be specifically templated using an array of nanoscale topographical elements that act as surrogates for the minority domains of the block copolymer. This technique has been demonstrated by several authors and is also called guided self-assembly or graphoepitaxy [98–104]. Although it is a nanopatterning technique that is straightforward and fairly simple to use, and has a large variety of design parameters, the difficulty in controlling long-range order and the defects experienced are a major challenge to applying this technique in real-world devices [105]. We also investigated guided self-assembly at the MESA+ Institute for Nanotechnology utilizing block copolymer, nanocrystals and supramolecular systems deposited onto templates fabricated by lithography, such as LIL. This topic will be discussed further, in section 4.1.10.

Confined synthesis is the final class of these novel templated techniques to be mentioned, Fan et al. demonstrated the hierarchical nanofabrication of microporous crystals with ordered mesoporosity, realized through confined growth within a three-dimensionally ordered mesoporous carbon, synthesized by replication of colloidal crystals composed of size-tunable silica beads of about 10–40 nm diameter [106]. Moreover, the authors revealed novel crystal morphologies, consisting of faceted crystal outgrowths from primary crystalline particles, thus providing new insight into constricted crystal growth mechanisms underlying confined synthesis.

4.1.9 Nanolithography in R&D

The formation of patterns by interference lithographic processes (IL) can be carried out with any type of coherent light source. However, IL becomes most interesting if a wavelength is used that reaches pattern periods which are below the resolution limit of an optical photolithographic aligner at $\lambda = 365$ nm. This section will describe some of the IL concepts, and presents a case study of this technique, which is often employed in research and development using a deep UV (DUV) laser with a Lloyd's mirror set-up to produce a cost-effective and stable laser interference lithography (LIL) exposure stage.

DUV Interference Nanolithography

Schattenburg et al. fabricated nanostructures in 1995 by utilizing an Argon-ion laser source at $\lambda = 351.1$ nm [107]. The authors suggested interferometric lithography was the tool of choice for patterning large-area, phase-coherent periodic structures with submicron-period gratings and grids, and they postulated many applications, including x-ray and extreme ultraviolet (EUV) transmission gratings, EUV-UV polarizers and filters, field emitter arrays for flat panel displays, as well as in reference grids for e-beam lithography.

An example of using this type of nanolithography of sub-micron sparse hole arrays for field-emission display applications is given by Chen et al. [108]. This paper presents the use of multiple exposure at two different pitches (i.e., angles of incidence), the integration of interferometric and optical lithography by Moiré alignment techniques, and various multi-beam techniques that provide a sparse array and result in a two-dimensional pattern in a single exposure. Later, the same group presents an overview, which considers interferometric lithography to be a nanotechnology enabler in a large variety of applications, but specifically for large-area samples [109]. Several further publications concerning advanced, maskless lithography for the writing of arbitrary non-periodic patterns have been presented by researchers from the Bell Labs (today belonging to Alcatel Lucent, USA) [110–112], and another, more recent example of using interference in combination with catalytic etching is given by Choi et al. [113].

EUV Interference Nanolithography

Solak discussed how this technique produces periodic nano-structures with resolution below 20 nm [114]. Earlier work by Choksi et al. introduced the concept of maskless extreme ultraviolet lithography [115], and Olav Solgaard and co-workers at Stanford University, USA, continued these developments, publishing work on analytical models for several micromirror array configurations and later on spatial light modulators for maskless lithography [116, 117]. More recent progress in extreme ultraviolet interferometric and holographic lithography is published in several contributions by Cerrina's group at the University of Wisconsin-Madison, USA [118–120].

Atom Lithography

Laser focusing of chromium atoms for nanostructure fabrication was first decribed by McClelland et al. [121]. This work was followed up by a publication on a fast, deterministic source of single chromium atoms [122]. So far, this topic is progressing very slowly in the scientific community, with just one additional publication concerning the principle of Stern–Gerlach interferometry [123].

4.1.10 LIL Development at MESA+ NanoLab NL

A Lloyd's mirror laser set-up utilizing a cost-effective 266 nm DUV laser has proved to be a simple and useful tool for the fabrication of nanoarrays which have structural resolution down to the sub-500 nm regime. Combinations of reactive ion etching can be explored using the LIL resist as an etch mask. LIL was previously demonstrated by researchers at the MESA+ for the generation of sieves, photonic crystal structures and patterned magnetic media (PM) [124–128]. Figure 4.7 shows the concept for LIL exposure, and an example of a resulting patterned media in the inset.

The NanoLab NL cleanroom facility at University of Twente's MESA+ Institute for Nanotechnology, Enschede, The Netherlands, has installed an optical table that is appropriate for conducting laser interference lithography with the ability to change the experimental conditions for a variety of different projects. Based on earlier research by Cheng et al., who investigated LIL templates for the fabrication of nanostructures with long-range order using block copolymer lithography, for example, the MESA+ set-up has also been applied for such hybrid lithographic techniques [129]. Dziomkina et al. have investigated the patterning of surfaces where LIL assists the symmetry control of polymer colloidal monolayers and crystals formed by electrophoretic deposition [130]. Similar processes utilizing LIL were also demonstrated by other groups, e.g., Kim et al. [131].

The MESA+ LIL system was used for preliminary templating experiments in the layer-by-layer deposition of supramolecular assemblies. However, LIL has now been replaced by nanoimprint lithography (NIL) techniques based on e-beam lithographically produced masters for research

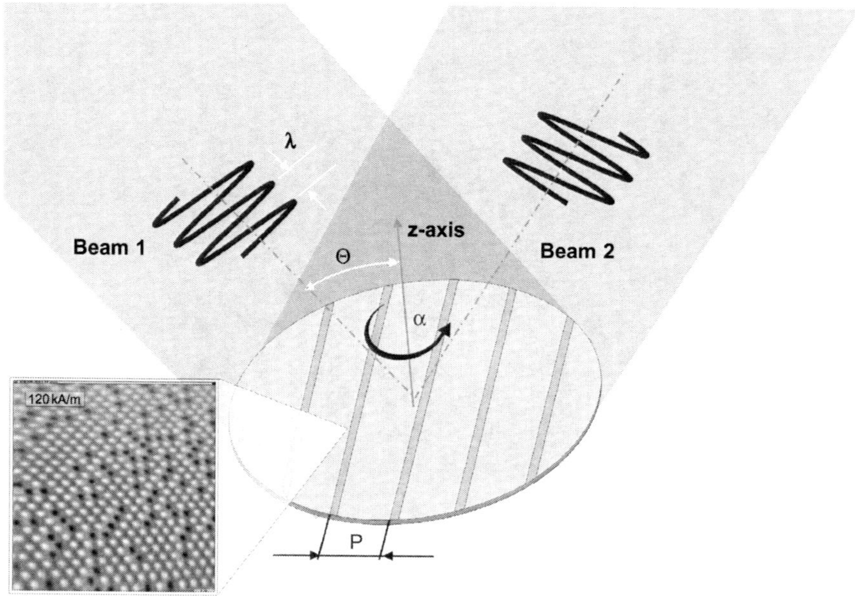

FIGURE 4.7

Artist impression of laser interference lithography for the fabrication of PM. The pattern period (P) is defined by the wavelength (λ) and the angle of incidence (Θ) while the possibility to rotate the substrate allows two-dimensional arrays such as dots, hexagones, and Moiré-type patterns through a multi-exposure at different substrate rotations (α). LIL has been applied to fabricate the magnetic structure shown here in the inset, which is described in detail by Murillo et al. [127]. (Reproduced from reference [2] with permission from IOP Publishing.)

into this hybrid approach [132]. Figure 4.8 shows the set-up as used at NanoLab NL. The optical set-up allows full scale 4-in. wafer handling.

A variety of current investigations of nanostructure applications continue to rely on LIL as a nanolithographic tool at MESA+, as is mentioned in the following section.

Multi-exposures and Novel Resist Systems for 266 nm LIL

This section presents some additional results that have been achieved in the nanofabrication of large-area nanostructured surfaces, and explores the capabilities of specific resists and multi-exposure approaches. Both method and materials are briefly summarized here.

FIGURE 4.8

NanoLab NL cleanroom facility at MESA+ Institute for Nanotechnology. Full scale 4-in. wafer handling at the LIL set-up. (Image courtesy: H.A.G.M. van Wolferen, University of Twente, 2011.)

Method/Materials

The generation of one-dimensional (gratings), two-dimensional (arrays of dots) and mesoscale quasi-regular discontinuous nanoarrays by LIL exposure into a resist can be described briefly as follows. A standard (100) p-type, polished prime silicon wafer is used as the substrate. LIL is operated at 266 nm, utilizing the aforementioned continuous-wave laser source as an exposure tool. This lithographic process line, consisting spin coating, soft-bake, LIL exposure, post-exposure-bake and wet development, is described in detail earlier [133]. Three types of resist were employed to evaluate their performance in the LIL set-up. The first, a standard i-line photolithographic resist system OiR 907-13 (from Fujifilm, formerly called Olin) was diluted to a suitable concentration to be spin-coated and produce a film thickness below 200 nm. The use of diluted Olin for 266 nm-LIL has also been described by Murillo et al. [127]. Although the resist is not very stable after dilution, we applied this resist system for multi-exposure experiments, producing a mesoscale quasi-regular nanoarray (for an overview of nanoarray definitions see Figure 4.2 and Figure 4.3).

The second set of conditions for producing nanogratings by a single exposure used a positive chemical amplified resist, PEK500 (Sumitomo Chemical). This was set up for a systematic evaluation of scaling line width, pattern period (pitch) and pattern depth (aspect ratio) during pattern transfer by RIE. A full record of this fabrication technology exploring positive chemical-amplified resist (p-CAR) has been given previously [133]. Finally, a third type of resist, the negative acting MA-N2403 (Microresist) demonstrated the potential for using high-contrast resists during LIL multi-exposures to generate integrated micro-nanoarrays.

Results

Multiple exposure experiments at two different angles of incidence (14° and 15.4°) and repetition of these exposures after a substrate rotation of 90° were carried out to achieve a latent nanopattern grey-scale image in the OiR photoresist, here acting as a low-contrast resist at $\lambda = 266\,\mathrm{nm}$ to generate Moiré-type features. Subsequently, the resist was developed in a single step in an immersion bath for 1 min using standard OPD4262 positive resist developer. Figure 4.9 shows how these structures evolved from interference between the multiple set of exposures.

Figure 4.10 demonstrates the potential of p-CAR pattern transfer by parallel plate reactive ion for etching a dot pattern into a layer of silicon

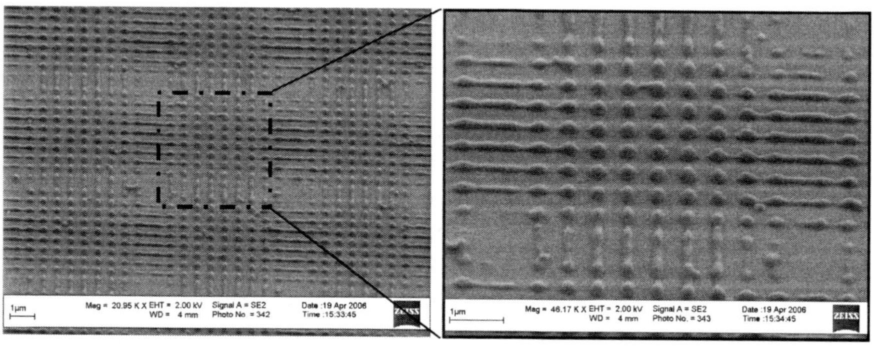

FIGURE 4.9

Four multiple exposures resulting in a spatially defined pattern of micrometer and nanometer sized elements within a quasi-regular array, previously also defined as a discontiuous nanoarray. The mesoscale pattern can be uniformly fabricated across an area of approximately $2 \times 2\ \mathrm{cm}^2$. (Image courtesy: M. Smithers, University of Twente, 2006.)

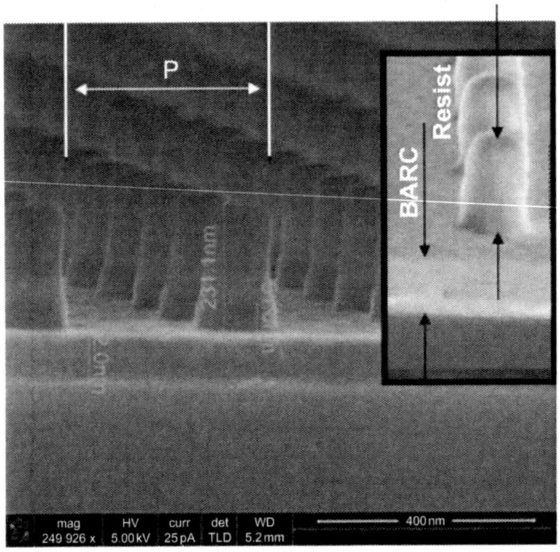

FIGURE 4.10

Dot array on silicon nitride. Orthogonal exposure into resist on bottom antireflection coating (BARC) transfer layer at 20.48° angle of incidence resulting in a periodic pattern in the resist with a pitch (P) of 380 nm (inset shows an example of the resist prior to the pattern transfer etch). The scanning electron micrography (SEM) shows a cross-sectional view of the silicon substrate at nearly 90° of the SEM holder producing a full-height view on the dots etched partly into the silicon nitride. The resist/BARC mask has not yet been removed. (Image courtesy: J.G.M. Sanderink, University of Twente, 2007.)

nitride, aiming at a pattern period (pitch) of 38 nm, which refers to an angle of incidence of 20.48° in the two orthogonal exposures. The dots can be subsequently coated with magnetic, conducting material. Previously, these nanoarrays were fabricated using an orthogonal exposure in OiR resist. Subsequently, the samples were used in a method for field-emission distance control by le Febre et al. [128]. Applying the improved LIL process utilizing p-CAR resist, as shown here, will further support well-controlled sample preparation for PM recording experiments.

Figure 4.11 shows the nanograting pattern fidelity of the line width/space ratio at five different points across the diameter of a 4-in. silicon wafer, as measured by scanning electron microscopy. This result demonstrates the difficulty of controlling the uniformity of such continuous nanoarrays over a large manufacturing area. This aspect remains a challenge for the exposure tool designer in selecting suitable optics.

When replacing the low-contrast OiR resist with a high-contrast negative resist, e.g., the MA-N2403 (Microresist) used here, a highly-defined

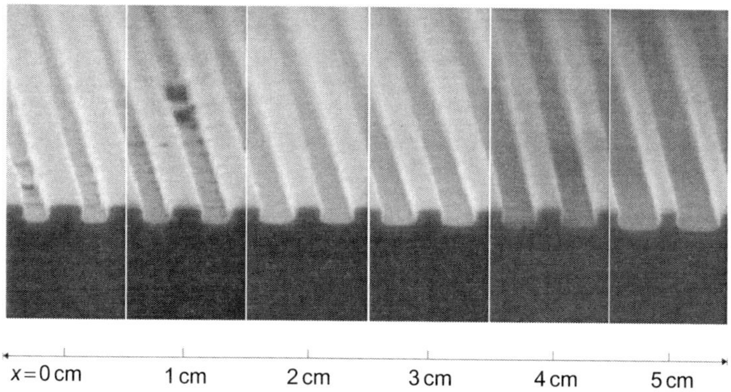

FIGURE 4.11

Single exposure at 40.6° angle of incidence [134]. The scanning electron micrographs are taken on the middle line of the wafer to demonstrate the pattern distortion due to optical aberrations. However, the usable area of high uniformity still stretches across a few cm^2 during the full 4-in. wafer scale exposure process. Areas of high pattern fidelity vary with the angle of incidence applied and the total mirror surface area being exposed uniformly by the source. (Image courtesy: M. Domanski, University of Twente, 2008. Reproduced from reference [2] with permission from IOP Publishing.)

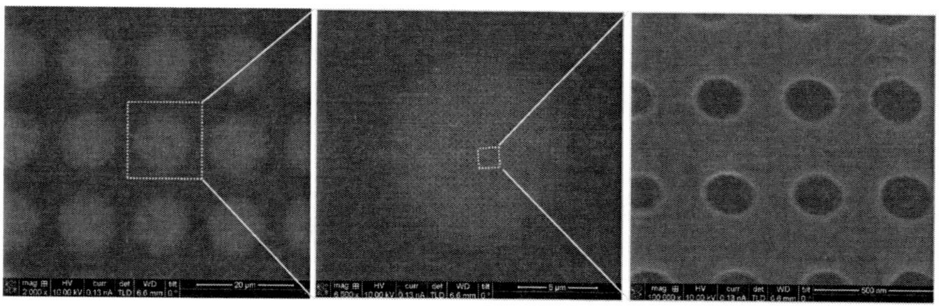

FIGURE 4.12

Nanohole-microarray formed by high-contrast negative resist LIL-multi-exposure producing nanoholes at 20° angle of incidence. Details of this processing conditions are disclosed. (Image courtesy: H.A.G.M. van Wolferen, University of Twente, 2009.)

integrated micro-nanoarray pattern results from the same type of multi-exposure carried out in the previous experiment (Figure 4.9). A micro-nanoarray produced by using the negative resist is presented in Figure 4.12.

4.1.11 Case Study: Laser Interference Lithography Nanoarrays for Cell Biology

Specific experiments have been designed to study the interaction of biological material (tissue, blood, proteins) with nanostructured surfaces. First, systematically scaled, large-area nanogratings were fabricated by solvent cast replication from the large-area (2×2 cm^2) master, by the LIL process and subsequent silicon etching. Secondly, these nanostructured surfaces were investigated for their cell response. The fabrication details are communicated elsewhere [92, 135], as are the full details of the biological experiment [92, 136]. Here, we only wish to stress the multitude of processing possibilities of LIL nanolithography.

Several studies have demonstrated that biological processes can be activated when a surface has a specific nanotopography [137, 138]. Kim et al., for example, cultured fibroblasts on PMMA surfaces that were nanostructured using AAO as a template for nanoimprinting [91]. Recent work by van Delft et al. showed that e-beam lithography can be applied to manufacture a template containing nanopatterned fields of $500 \times 500\,\mu$m^2 that were reproduced in polystyrene and subsequently tested for their cell response [139]. The lines and spaces of the nanopattern within each field had various pitches, but always at a line width-to-space ratio of 1:1. Original work relates back to micropatterned surfaces for cell alignment in cell culture [140].

Lamers et al. systematically investigated the size and geometrical dependence of the biological response in cell culture experiments in the context of tooth implants. LIL is used in this context for its capability for large-area nanoarray formation, and also its potential to be used for simple and cheap maskless exposure onto non-flat surfaces.

Large-area LIL nanoarrays with a diversity of line width, pattern period and depth have been evaluated for their fibroblast, osteoblast and microphage response [92]. Initial results concerning the fabrication process for artificial nanotopography in medical grade titanium utilizing LIL has been presented by Domanski et al. [141]. Large-area patterning supported by the LIL technique is of particular importance in gaining statistically relevant results from the cell culture proliferation experiments. Substrates produced by e-beam lithography are often too small in surface area, only allowing a few cells to attach to one specific pattern.

4.1.12 Concluding Remarks on Emerging Nanolithography

In lithographic processes the control of the transfer of a layout first into a masking layer (resist) and then to a functional material allows the creation of devices with an enormous variety of capabilities. Of course, electronics remains at the frontier of lithographic techniques, reaching the scale of a few nanometers. Other disciplines are catching up quickly and even bring their own lithographic techniques.

A broad choice of techniques exists, of which only a few will be selected to lead market innovation in electronics. EUV is probably the preferred technique. In research investigating nanoscale behaviour in specific (e.g., PMs for magnetic data storage) and niche manufacturing (e.g., optical devices), templated techniques are gaining in significance. Nanoimprint lithography is gradually replacing mainstream UV-photolithography. When it comes to nanoarrays, interference lithography has become a valuable addition to the range of techniques.

The ability of chemical and biological substrates to coordinate ordering at the molecular scale is also about to reach the level of control required in industry. However, this is not yet on the wafer scale which we are used to in commercial applications. In respect to further deepening our understanding of controlling device functions at nanoscale dimensions, research in the field of nanolithography will remain extremely diverse and may be mainly guided by which techniques are to hand as opposed to a choice by design. It is, therefore, of particular importance to develop new cost-to-lithographic performance models, which give the designer of novel devices some insight into this diversity. Yet the choice between a well-established but limited process line, e.g., as is found in microelectronics complementary metal-oxide semiconductors (CMOS), or a novel lithography that capitalizes on the advantage of delivering very unique devices' properties is not an easy one. The multitude of parameters involved in establishing a production line makes such choice difficult.

The next direct-write techniques, e.g., using focused ion beams, are being developed. Their extension to an out-of-plane dimension may be either supported by stacking (assembly), multilayer processing, folding, or three-dimensional lithographic techniques such as two-photon lithography [142]. However, so far only a few technologies have shown true potential for creating three-dimensional crystal lattices across the different length scales from the nanometer to the centimeter, one of which is holographic lithography [143]. Whether these processes will find real application beyond photonic crystals is not yet known. The patterning of nanoarrays, e.g., for multilayers which are useful to perpendicular magnetic recording, x-ray diffraction, EUV diffusers and SAW devices, as well as surface modification by nanostructuring of implant materials, is, however, being actively investigated in our laboratory at the MESA+.

4.2 NANOMATERIALS

The integration of structural elements using lithographic patterning techniques aims for a more complex apparatus, which can be at various length scales, for example hand-held equipment. However, as mentioned above, pillars, holes or wires that are fabricated at the nanoscale have already shown unique properties, and ordering these in specific arrangements results in novel phenomena. Such materials are called metamaterials, an example of which is a material with a negative refractive index, creating an artificial structure. These exciting material properties bring about new opportunities for applications.

A functional device or system which demands some level of ordering in its materials also requires a carefully designed manufacturing process. Here, we will present an overview of nanolithographic techniques for oxides. Nanoscale designed oxides show great promise for energy storage and energy converting systems, and are thus a very important class of novel materials. Bio-inspired, templated oxides will be described in relation to other nanolithographic techniques, as will the way in which (nano)structured materials can deliver specific functionality at the interface with biological materials. We can look for new applications for these materials including within medicine, where the properties of oxide nanostructures as a biocompatible material are explored. Photonic crystals, for example, can be applied in medical diagnostic devices which are used in direct contact with complex biological samples, e.g., blood. In this section, oxide metamaterials are introduced and the emerging technologies which are employed for their fabrication, together with the medical devices that utilize oxide nanoarrays, will be discussed.

4.2.1 Ordered Oxides

It is the objective of this section to review nanolithographic techniques for producing oxide nanoarrays, specifically inorganic oxides. Although many interesting (polymeric) organic oxide arrays exist and are being investigated for medical applications, for example, thiolated polyethylene oxide as a non-fouling element for nano-patterned biodevices, we will not be able to discuss these here [144]. Some examples have already been given, such as nanostructuring of a polymer that is used as a colloidal masking layer (nanosphere lithography, see Figure 4.4D), where polystyrene beads protect the polymer film in a subsequently applied O_2 plasma etch process. Initially these polymerical films may be deposited by plasma enhanced chemical vapor deposition (PECVD) [145]. Another example is a highly experimental technique that utilizes nanobubbles as templates for the fabrication of nanoporous films of polypyrrole (PPy) on highly ordered pyrolytic graphite (HOPG) [146]. PPy is widely used as a conductive polymer in medicine, e.g., PPy-implants are suitable for neural prosthetics [147]. This text will not discuss oxide nanoarrays from a materials engineering point of view, although this has an extensive history of making textured materials through chemical synthesis. This book will, however, reflect on these systems from the viewpoint of a device designer who is screening the emerging nanotechnologies for nanostructured oxide devices. Therefore, we will first introduce the terminology as used in this context and then present some background on naturally occurring oxides. Our discussion will consider how metamaterials complement and enhance device functionality, making particular reference to the impact of nanoscale-provoked functionality in innovative devices in nanomedical applications.

4.2.2 Oxide Nanoarrays: Definitions and Background

Natural versus Artificial Oxide Nanoarrays

Before discussing the various oxide materials that form nanoarrays, we must first define what is meant by the term "oxide" in this context. Turning to inorganic chemistry, an oxide is a chemical compound formed from an element and at least one oxygen atom. Materials can also be classified by their oxidation state (the number of valence electrons in the chemical compound). Nearly all elements can form oxides when they are exposed to oxygen. Noble metals such as gold or platinum form oxides with greater difficulty, but these do exist. Oxides can be also gaseous, but we will only consider solid oxides.

Oxides can be amorphous (for example window glass) or crystalline (quartz, sapphire). Natural silicon oxide formed on a silicon surface is also amorphous, and is normally removed from wafers prior to processing. In this context, there is no need to explain the dielectric properties of silica and its benefits or disadvantages in electronic applications. In other applications, amorphous oxides have been used as a platform for releasing molecules or electrons from its matrix. For example, in drug delivery, highly defined oxides have replaced the polymeric coatings which were previously used. Such a polymer-free platform is used for drug loading in cardiovascular stents [148]. Additionally, biomedical devices for neural stimulation have been based on oxide thin-film technology. In this application, a Teflon-type substrate is covered with a gold/iridium oxide film. The gold layer acts as current collector, and the iridium oxide as an electrochemical actuator. The intrinsic material microstructure plays an important role in enabling a low-resistance interface [149]. Another more recent example for an application of iridium oxides (IrOx) is given by Huang et al., who developed an IrOx micro pH sensor array [150].

Such miniaturized sensors could be used in chemical, food and bio-material devices. The authors present the design and fabrication process of an amorphous, IrOx, thin film, pH sensor array on flexible polymer substrates in their paper, so the utilization of amorphous oxides as micro- and mesoporous materials have therefore already found their way into applications. Most likely these materials will not be called nanoarrays, although their intrinsic microstructure means that they could be considered as such (e.g., see Figure 3.7, Chapter 3, Section 3.3.6), as could natural crystalline oxides. The properties and classifications of oxides is too broad a field to be discussed here, so this section will be restricted to some oxides that have been described in the literature as an artificially manufactured material. Examples are selected on the basis of their specific optical properties and the application of such oxide nanoarrays in the medical field. Figure 4.13 depicts the four different material types that can also be manipulated into nanoarrays: (A) crystalline, (B) polycrystalline, (C) amorphous and (D) layered composites.

Oxide Metamaterials: How Do They Differ from Nanoarrays?

One could argue whether the term *nanoarray* is really required. Both artificially formed and natural materials show ordering. At the nanoscale, many materials show a certain level of order and with respect to engineering, one likes to give the new field of research that investigates these phenomena a new name: nanotechnology. Hence the arrays being investigated at this scale are called nanoarrays. Most are formed by engineered processes and therefore

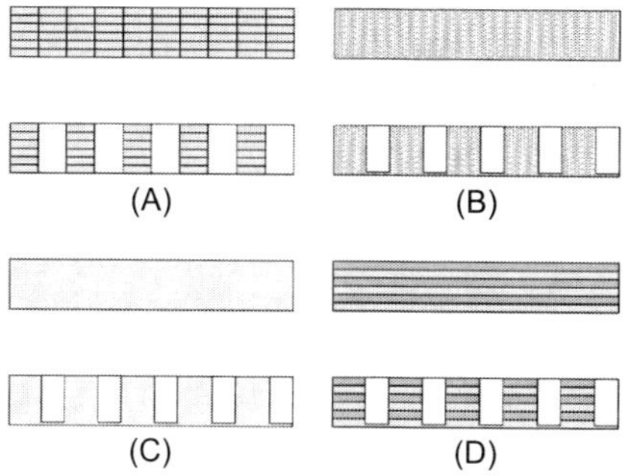

FIGURE 4.13

Definition of nanoarrays: (A) crystalline, (B) polycrystalline or interdispersed composite, (C) amorphous, and (D) layered composite materials and their according nanoarray by lithographic patterning. (Reproduced from reference [3]. Copyright © 2010 SPIE.)

these materials may exhibit some property that differs from naturally occurring structures, e.g., structures with a negative refractive index. However, since these artificial materials are generally known as called metamaterials, their properties may not always be derived from the nanoscale. The term *nanoarray* is most often used for nanoscale engineered metamaterials which contain a high ordering of a nanoscale object, such as a pillar or a hole. This is a very exciting class of materials. There is also a wide variety of such materials that are grown epitaxially (gaining their order from the underlying crystalline substrate). After a short general overview given in the next two subsections, the core of this section discusses nanoarrays from oxides that are formed by manipulation of material by a lithographic-type process, and have subsequently demonstrated exceptional material properties.

Curiosity-Driven Research on Nanoscale Metamaterials

A general theory of electromagnetic propagation in periodically stratified media has been proposed by Yeh et al. [151]. It was found that point and line defects could be introduced into the structure to localize and direct optical waves. Another article describing the optical properties of nanostructured optical materials was published by Gehr and Boyd [152]. These are just two publications of the many which discuss these phenomena. When going deeper into this topic one should realize that all applications start here. Industry will probably not invest directly in research into the fundamental behavior of matter at an energy boundary encompassed by the material's properties. Fortunately, questions concerning these underlying working principles and rules of nature can be investigated by independent and curiosity-driven scientists, due to public funding, if this continues.

Application-Driven Research on Nanoscale Oxide Metamaterials

Enhancing the optical properties of integrated sensing devices is an interesting research topic. Rhenium oxide films have been characterized by Cazzanelli et al. [153], and the fabrication and characterization of cerium oxide (CeO_2) thin films for ultraviolet sensing applications was recently presented by Wang et al. [154]. The latter use oxide deposition by reactive oxygen ion-beam assisted e-beam evaporation to produce cerium oxides with refractive index enhanced from ca. 1.8 to 2.2. This ion-assisted deposition method also increases electrical resistivity by a factor of ten. This was a new concept for photodiode devices. A gain in photocurrent of 100 times and a fast photoresponse under a 405 nm UV exposure had been discovered. These parameters, probably both linked to the material's specific stoichiometry and crystal structure, are measurable by x-ray photoelectron spectroscopy and x-ray diffraction.

Pulsed laser deposition (PLD) is an emerging material deposition method. Nanoporous cerium oxide thin films were desposited by PLD and used for glucose biosensors. The sensors containing such films showed enhanced enzyme affinity for converting immobilized glucose oxidase (GOx) to glucose, as reported by Saha et al. [155]. They studied the surface morphology of these PLD CeO_2 films by atomic force microscopy. Their results show that CeO_2 nanoporous films can be utilized in glucose sensing applications without the need for additional surface functionalization. Another example of this type of sensor was presented by Ansari et al. [156], in a feasibility study of glucose oxidase (GOx) immobilized tin oxide thin films, consisting of nano-baskets, for glucose sensing. Their approach used anodized alumina pores of 80 nm diameter as a template for the plasma enhanced chemical vapor deposition (PECVD) of tin oxide (SnO_2). This process formed nano-baskets. The GOx was immobilized by physical adsorption from solution, and elemental analysis of the films revealed the GOx was covalently attached to the tin oxide. This sensor design, based on a crystalline film with basket-like (nano-sized) structures, demonstrated higher sensitivity in comparison to other structures.

Investigating conducting metal oxides, Rhodes et al. reported initial observations of surface plasmon resonance (SPR) [157]. They claimed that the systematic variability of many conducting metal oxide composites suggests significantly advanced film properties and innovative possibilities for versatile and sensitive chemical sensing applications. The properties of nanoscale oxide metamaterials and their fabrication processes could potentially be utilized in devices other than sensors. In 2004, Vayssieres described a novel growth concept, and a low-temperature aqueous chemical growth thin-film processing technique that may lead to a new generation of smart and functional metal oxide particulate thin films and coatings, called purpose-built materials [158].

4.2.3 Principles of Oxide Nanoarray Fabrication

In Chapter 3 we referred to the patterning of oxides such as ceramics, zirconia or alumina, as well as more complex oxide composites that are partly or fully crystalline. These materials are difficult to pattern by etching techniques. Normally, ceramics are put into their desired shapes prior to sintering by ceramic-forming techniques such as slip-casting. In the top-down approach, thin-film oxides were nearly always patterned by physical ion bombardment, i.e., ion milling techniques, to create an artificial structure. This method is very inefficient, and more recently chemically-assisted reactive ion etching through appropriate masks, together with novel templating techniques, including lithographically assisted templating, have been under development (e.g., MIMIC, see Chapter 3, Section 3.3.3). Because ceramics are difficult to manufacture from the top-down, most research attention has been given to the design of morphology through a materials engineering approach, which aims to control the intrinsic microstructure from the bottom up. Some examples of nanoporous sensor designs have been given already. Pre-1995, the bottom-up approach in this field would have been treated as coordination chemistry, and would probably have used metal-ion coordination either by a wet-chemical technique or from a gaseous precursor [159]. However, to generate metamaterials with more engineering versatility, recent advances in deposition techniques such as pulsed laser deposition (PLD) present the possibility of stoichiometrically defined materials from ceramic targets. These targets are often prepared from pressed powders or by ceramic slip casting. PLD allows the deposition of highly innovative multilayer materials. Ravichandran et al., for example, contributed to the understanding of the nature of electronic effective mass in double-doped $SrTiO_3$ by this technique [160].

Sol-gel, and electrochemical processes (such as the anodic oxidation of aluminum) can also be used to form highly ordered oxide films with specific morphology. Much research is currently being devoted to investigating techniques for patterning oxide films at the nanoscale. Some studies discuss the intrinsic modification of crystal structure through laser radiation. Beke et al., for example, investigated laser treated vanadium oxide sols [161, 162]. Their x-ray photoelectron spectroscopy and x-ray diffraction studies reveal how the original xerogel particle structure changes into irregularly shaped, layer-structured V_2O_5. Depending on the laser power used, different material structure was created. Some specific examples of non-lithographic templating have already been mentioned in section 4.1.8. The following sub-sections will focus on top-down lithographically-assisted, collodial, bio-inspired and biotemplated techniques for oxide nanoarrays.

Lithographic-Assisted Oxide Nanoarrays

In this class of patterning techniques two controversial approaches have been discussed for many years. These are the direct-write lithography approach, which is normally used in industry for the primary patterning of a mask, and lithography incorporating the pattern transfer through a mask. There is a wide range of available methods, and the question is which of them could be applied to creating oxide nanoarrays. So-called electro-pen and dip-pen nanolithographic techniques, for example, are likely candidates [163, 164]. These nanochemistry and scanning probe lithographies were reviewed by Garcia et al. [165] among others.

I will restrict this discussion to two examples of very controversial concepts: oxide nanoarrays etched through a large-area mask, and focused ion beam (FIB) milling for prototyping integrated optical devices from novel materials.

An example of the first concept is the fabrication of nanoporous silica films by utilizing a chromium pattern transfer etch mask formed by particle lithography [166]. FIB, the second concept, has been applied by Lin et al. to experimentally confirm their simulations for two-dimensional nonlinear photonic crystals using barium titanate thin films [167]. Lee et al. have also used this technique to measure the Brillouin gain-coefficient for a bismuth-oxide-based photonic crystal fiber under significant beam reflection at splicing points [168]. Unfortunately, there are no specific industrial applications to date of these lithographic-assisted oxide nanoarray systems on the market.

Bio-inspired and Biotemplated Oxide Nanoarrays

Even further from industrial applications are the biomimetic approaches to material synthesis, as for example, reported by de Guire et al. [169]. Oxide thin films are deposited on organic self-assembled monolayers (SAM) from solutions at temperatures below 100°C. Dense thin films of ZnO were achieved under basic conditions. TiO_2, ZrO_2 and Y_2O_3, which were formed from acidic solutions, produced better films. Taking the technique a step further, this SAM layer can be pre-patterend by lithographic techniques such as nanoimprinting. However, direct write approaches and laser interferences may also be used in combinations with SAMs; for example, see Yasseri et al. or Friebel et al. [170, 171]. Alternatively, dip-pen lithography may be used to deposite SAMs or sol-gels locally onto a surface in a similar way to that described for matrix-assisted dip-pen nanolithography and polymer-pen lithography by Huang et al. [150]. A combination of the two processes (biotemplating and lithography) may present us with a design route for an even higher hierachical level of structural components than currently exists in microelectronics. Another approach to templating is the use of biological material such as aminoacids, peptides, DNA, microtubules, silk, etc. Viruses or bacteria have also been exploited as non-planar templates, e.g., for the formation of advanced, virus-free nanoparticles demonstrating virus-templated silica nanoparticles through mineralizaton of a cage-like viral bionanoparticle, such as the cowpea mosaic virus [172].

These systems have a natural tendency to form highly ordered features that can be utilized for the artifical patterning of engineered materials. Han et al., for example, utilized natural templates such as peacock feathers for novel photonic crystals that incorporate Nano-CdS, thus producing novel hybrid photonic crystal materials [173]. Affiliated with this research group, Su et al. presented the biogenic synthesis of highly ordered oxides achieved by stepwise bioredox/artificial oxygenation using egg-shell membranes (ESM) as a template [174]. Another recent example is the work of Zhang et al., who presented a novel photoanode structure templated from butterfly wing scales to embed zinc oxide (ZnO) nanoparticles, replicating the naturally occurring nanostructure in these systems by in-situ ZnO nucleation at suitable binding sides on the template [175].

Although access to state-of-the-art analytical equipment, such as high resolution scanning electron microscopy, is necessary in this field of research, these contributions clearly demonstrate that innovation through nanotechnology research does not always need large capital resources. Any of these template techniques have the goal of producing functional hybrid nanocomposites with hierarchical micro-nanostructure. Biomimetics of photonic nanostructures were introduced as an engineering approach by the Parker and Townley methodology [176]. See also the review "Natural photonics for industrial inspiration" [177]. A study of biomimetic interfaces with high-performance optics in the deep-UV light range has been performed Lohmüller et al. [178], and Balci et al. demonstrated the catalytic coating of virus particles with zinc oxide utilizing a rod-shaped tobacco mosaic virus. These electrodeless deposited materials show cathodoluminescence [179]. The generation of oxide nanopatterns by combining the self-assembly of S-layer proteins and the area-selective atomic layer deposition of metal-oxide-based high-k materials was demonstrated by Liu et al. [180]. They investigated the deposition of hafnium oxide in the pores of the octadecyltrichlorosilane (ODTS)-modified S-layer proteins on silicon. These materials will facilitate further miniaturization of logic and memory nanoelectronic devices.

Many groups are currently exploring biology and biomolecular engineering for potential technical benefits to nanofabrication. A variety of these biomolecular assembly strategies have demonstrated groundbreaking results [181]. The developments of these novel bio-hybrid concepts also indicate the possibility of a new era of highly sensitive, low-cost diagnostic devices and efficient novel medical therapies.

4.2.4 Ordered Oxides in Medical Applications

Actually, nanofeatures in metal exhibiting exceptional optical behavior are already much more common than nanostructures in oxides. Experimental and simulated near-field images of a nanohole array in gold films has been reported by Gao et al. [182], who give "Direct evidence for surface plasmon-mediated enhanced light transmission though metallic nanohole arrays". One would like to argue that it is the actual hole structure which causes the enhanced light transmission, and not the metal. This hypothesis has been tested experimentally by creating the same structural features in silicon, which did not exhibit the same effect. Therefore, the combination of the material and the nanostructuring led to the desired effect. Such type of designs were also characterized by a prominent diffusion phenomenon in the near ultraviolet region [183]. Many oxides demonstrate exceptional optical behavior, such as the anomalous enhancement of Raman scattering that was detected for tungsten oxide thin films deposited on indium tin oxide (ITO) coated glasses during special annealing treatment [184].

Novel Nanoarrays in Diagnostics

The fabrication of inexpensive but highly accurate and sensitive diagnostic devices is dependent on the development of novel fabrication technology. Nanotechnology moves diagnostics from microarrays towards nanoarrays for high-throughput analysis. Further downscaling would be advantageous for point-of-care application and field applications [185].

By indirectly exploiting intrinsically confined oxide nanoarrays, scientists at the University of Illinois, USA, have developed a disposable, microplate-based optical biosensor that is capable of detecting protein-DNA interactions. The photonic crystals biosensor consists a hybrid material. First a low-refractive-index polymer grating is manufactured and coated with high-refractive-index titanium oxide. The grating structure is positioned on the bottom of a standard 384-well microplate, in which each well functions as a miniaturized test tube [186]. In the light of this paper, protein nanopatterns have been utilized to improve immunodetection sensitivity. The sensor surface is prepared with bioadhesive domains of nanometric dimensions in a nonadhesive matrix which is formed by means of colloidal lithography [187].

Perspectives on Nanoarray in Therapy

Engineered nanomaterials for biophotonic applications to improve sensing, imaging and therapeutics are clearly feasible [188], but there seems to be no specific examples of the use of oxide nanoarrays in this regard in the literature. When referring to the use of organized systems there may be room for improvement in photodynamic therapies. Furthermore, we could modulate the release rate of a therapeutic agent from a membrane (or a vesicle) by tuning its pore size, e.g., by modulation of these properties by means of optoacoustic or optoelectronic phenomena. By using such smart biomaterials, for example as skin-scaffolds, photo-stimulation may support the wound-healing process. More realistic at this stage are optical means of material functionalization prior to contact with biological material. Future applications may include opportunities in scaffolds for cell-growth, substrates for cell culture, experiments in biomedical-related diffractive optics and optical interfaces, and substrates for tensegrity studies. Perry et al. gives an example for nano- and micropatterning of optically transparent, mechanically robust, biocompatible silk fibroin films [189].

4.3 WHERE ARE WE?

Trillions of nanosystems are packaged with ease in our body. Nature has mastered the assembly of nanosystems into workable machines through evolution. In the engineering sciences we can learn from nature, and use our findings as a source of inspiration for packaging artificial, higher assembled architecture devices. Of course, there is no competition between nature and engineering – it is synergy. Sometimes applying the concepts of nature and building from the bottom up would simply take too much time and the top-down approach works better. In the next decade it is important that the engineering sciences deliver appropriate design rules for industry to justify investments in nanotechnology. This survey of design rules for nanoengineering must, and will be, clearly supported by governmental incentives for creative research and development in this field. Based on such research incentives the investment risk for industries can be further lowered, specifically in areas which have long development cycles, such as medicine and high-tech applications. Currently there are no clear solutions for novel products that require first appropriate packaging of nanodevices as such that they can interact within real world devices.

4.4 WHERE TO GO FROM HERE?

In the past decade more and more developments have gone *nano*, and research on this topic has clearly intensified in many working fields. Nano has made a profound impact on medicine and health care industries, such as biomedical diagnostics. Investments in nano already presents us with novel opportunities in these branches of science, although products have not yet emerged on the market. It has been the objective of this chapter to reflect and highlight current trends of nanofabrication, and identify key issues and challenges for making nanomaterials with a high structural order – the so-called nanoarrays. We have covered several aspects of the fabrication of such metamaterials.

The implementation of novel nanofabrication strategies in medicine will, however, strongly depend on appropriate medical technology assessment protocols which incorporate materials screening through a multidisciplinary approach. New materials discovered in the Natural and Engineering sciences often take more than 10 years to appear in Life Sciences applications. Materials in contact with the human body are currently benefitting the most from these innovative nanosolutions. However, at least three aspects of such new developments in medicine must be investigated, besides their direct functional purpose. They are:

1. long-term safety in production and use of the nanomaterials,
2. match with a patient's personal profile, and
3. functionality throughout a timescale that meets the patient's needs in a sustained and reliable manner.

Healthcare professionals will only implement a new material in medical practice if these criteria are fulfilled. It is therefore an engineer's responsibility to create boundary conditions that will allow the assessment of a material's usefulness for this application as early as possible. Multiscale physics, exploiting a wide variety of advanced computational and experimental tools, is one approach to faster materials implementation and aims at a better understanding, a better description and – in medicine – a better control of the properties of these materials in biological processes. It allows the prediction of material properties or system behavior based on a knowledge of the atomic structure and properties of elementary processes. Photocatalytic activity of titania nanohole arrays has been demonstrated, for example, by Hamaguchi et al. [190], and PLD on colloidal templates has been evaluated for its ability to produce photocatalytic active amorphous TiO_2 nanocolumn arrays [191]. These titanium nanoarrays can be deposited on nearly any substrate and although they have only so far been discussed for possible use as a self-cleaning surface, such a technique may also foster medical applications. Healing, tissue regeneration and tissue engineering may directly benefit from these developments for inflammation-supressing scaffolds or dressings.

These nanoarrays need a large spectrum of fabrication techniques, and input from chemical and materials engineering sciences is still required in order to be able to broaden their field of application. However, meta-materials designed at the nanoscale are not yet easily accessible and robust production techniques have not yet been developed. Currently, most research effort is directed towards the better understanding of material properties, and only a small number of applications make use of these materials in biosensors. Nanotechnological applications are in their infancy [192]. In sum, research and developments in this field will benefit from a multidisciplinary approach which merges theoretical and experimental efforts. Related to the latter, a purpose-driven nanoengineering sciences is required to help in lowering the risk of industrial implementations, which will eventually foster overall greener production and use of resources.

References

[1] Z. Cui, *Micro-Nanofabrication*, Springer, Berlin, 2005, ISBN 7-04-017663-7.

[2] R. Luttge, Massively parallel fabrication of repetitive nanostructures: Nanolithography for nanoarrays, *J. Phys. D Appl. Phys.* **42** (12) (2009) Art. no. 123001.

[3] R. Luttge, Nanolithography for oxide nanoarrays and their application in medical devices, in: *Proceedings of SPIE – The International Society for Optical Engineering*, vol. 7603, SPIE, Bellingham, WA, USA, 2010.

[4] J.V. Barth, G. Costantini, K. Kern, Engineering atomic and molecular nanostructures at surfaces, *Nature* **437** (7059) (2005) 671–679.

[5] P. Havu, V. Havu, M.J. Puska, M.H. Hakala, A.S. Foster, R.M. Nieminen, Finite-element implementation for electron transport in nanostructures, *J. Chem. Phys.* **124** (5) (2006) 054707.

[6] A.O. Adeyeye, N. Singh, Large area patterned magnetic nanostructures, *J. Phys. D Appl. Phys.* **41** (15) (2008) 1–29.

[7] H.R.C. Dietrich, J. Knoll, L.R. van den Doel, G.W.K. van Dedem, P.A.S. Daran-Lapujade, L.J. van Vliet, et al., Nanoarrays: A method for performing enzymatic assays, *Anal. Chem.* **76** (14) (2004) 4112–4117.

[8] R. Ji, M. Hornung, M.A. Verschuuren, R. van de Laar, J. van Eekelen, U. Plachetka, et al., UV enhanced substrate conformal imprint lithography (UV-SCIL) technique for photonic crystals patterning in LED manufacturing, *Microelectron. Eng.* **87** (5–8) (2010) 963–967.

[9] P. Rai-Choudhury (Ed.,) *Handbook of Microlithography, Micromachining, & Microfabrication*, vol. 1, SPIE Optical Engineering Press, Washington, USA, 1997, ISBN 0-8529-6906-6.

[10] International Technology RoadMap for Semiconductors. 2007 Edition. http://www.itrs.net/Links/2007ITRS/2007_Chapters/2007_Lithography.pdf, 2007 (accessed 06.04.2011).

[11] Mapper Lithography. www.mapperlithography.com (accessed 06.04.2011).

[12] F. Hua, J. Shi, Y. Lvov, T. Cui, Patterning of layer-by-layer self-assembled multiple types of nanoparticle thin films by lithographic technique, *Nano Lett.* **2** (11) (2002) 1219–1222.

[13] K.-B. Lee, J.-H. Lim, C.A. Mirkin, Protein nanostructures formed via direct-write dip-pen nanolithography, *J. Am. Chem. Soc.* **125** (19) (2003) 5588–5589.

[14] F. Huo, Z. Zheng, G. Zheng, L.R. Giam, H. Zhang, C.A. Mirkin, Polymer pen lithography, *Science* **321** (5896) (2008) 1658–1660.
[15] S. Sekula, J. Fuchs, S. Weg-Remers, P. Nagel, S. Schuppler, J. Fragala, et al., Multiplexed lipid dip-pen nanolithography on subcellular scales for the templating of functional proteins and cell culture, *Small* **4** (10) (2008) 1785–1793.
[16] K. Salaita, Y. Wang, C.A. Mirkin, Applications of dip-pen nanolithography, *Nat. Nanotechnol.* **2** (3) (2007) 145–155.
[17] J. Bokor, A.R. Neureuther, W.G. Oldham, Advanced lithography for ULSI, *IEEE MCD* **12** (1) (1996) 11–15.
[18] J. Bokor, Directions in advanced IC lithography: New applications for lasers and coherent sources, in: *Summaries of Papers Presented at the Quantum Electronics and Laser Science Conference QELS '96*, 2–7 June 1996, Anaheim, CA, USA, 1996, pp. 27–28.
[19] R.M. Brown, Z. Barnes, C. Sawatari, T. Kondo, Polymer manipulation and nanofabrication in real time using transmission electron microscopy, *Biomacromolecules* **8** (1) (2007) 70–76.
[20] M. D'Alessandria, A. Lasagni, F. Mücklich, Direct micro-patterning of aluminum substrates via laser interference metallurgy, *Appl. Surf. Sci.* **255** (5 PART 2) (2008) 3210–3216.
[21] P. Argitis, D. Niakoula, A.M. Douvas, E. Gogolides, I. Raptis, V.P. Vidali, et al., Materials for lithography in the nanoscale, *Int. J. Nanotechnol.* **6** (1–2) (2009) 71–87.
[22] R. Schenker, W. Oldham, Damage-limited lifetime of 193-nm lithography tools as a function of system variables, *Appl. Opt.* **37** (4) (1998) 733–738.
[23] M. Van De Kerkhof, W. De Boeij, H. Kok, M. Silova, J. Baselmans, M. Hemerik, Full optical column characterization of DUV lithographic projection tools, in: *Proceedings of SPIE – The International Society for Optical Engineering*, vol. 5377, SPIE, Bellingham, WA, USA, 2004, pp. 1960–1970. Part 3.
[24] P.D. Harris, Practical DUV lithography for the optoelectronics market, *Microlith. World* **13** (1) (2004) 18–20.
[25] M. Okumura, J. Ishikawa, M. Hamatani, M. Nei, Mass production level ArF immersion exposure tool, in: *Proceedings of SPIE – The International Society for Optical Engineering*, vol. 6154 II, SPIE, Bellingham, WA, USA, 2006. Part 2.
[26] H.H. Magoon, Scanner influences on resolution capabilities, in: *ASMC (Advanced Semiconductor Manufacturing Conference) Proceedings*, 2005, pp. 175–179.
[27] W.-H. Cheng, K. Chakravorty, J. Farnsworth, Alternating phase shift mask architecture scalability, implementations and applications for 90 nm & 65 nm technology nodes and beyond, in: *Proceedings of SPIE – The International Society for Optical Engineering*, vol. 5130, SPIE, Bellingham, WA, USA, 2003, pp. 766–777.
[28] L. Collins, Moving on from Moore [miniaturization], *Eng. Technol.* **2** (1) (2007) 26–27.
[29] C.-F. Wang, S.-F. Chiou, F.-H. Ko, J.-K. Chen, C.-T. Chou, C.-F. Huang, et al., Polybenzoxazine as a mold-release agent for nanoimprint lithography, *Langmuir* **23** (11) (2007) 5868–5871.
[30] A.M. Ionescu, New functionality and ultra low power: Key opportunities for post-CMOS era, in: *Proceedings of Technical Program, In 2008 International Symposium on VLSI Technology, Systems and Applications (VLSI-TSA)*, IEEE Service Center, 445 Hoes Lane, Po Box 1331, Piscataway, NJ, 2008, pp. 72–73.

[31] J.P. Kent, J. Prasad, Microelectronics for the real world: "Moore" versus "More than Moore", in: *Proceedings of the Custom Integrated Circuits Conference*, CICC 2008, IEEE, Service Center, Piscataway, NJ, USA, 2008, pp. 395–402.
[32] R. Yu, High density 3D integration, in: *Proceedings, International Conference on Electronic Packaging Technology and High Density Packaging*, ICEPT-HDP, 2008.
[33] K. Chakrabarty, Digital microfluidic biochips: A vision for functional diversity and more than Moore, in: *Proceedings – IEEE Annual Symposium on VLSI*, ISVLSI 2010, Service Center, Piscataway, NJ, USA, 2010, pp. 3–4.
[34] K.D. Wise, Microelectronics in the "More than Moore" era, in: *Device Research Conference – Conference Digest*, DRC, 2010, Service Center, Piscataway, NJ, USA pp. 3–4.
[35] P. Ramm, A. Klumpp, J. Weber, M.M.V. Taklo, 3D system-on-chip technologies for More than Moore systems, *Microsyst. Technol.* **16** (7) (2010) 1051–1055.
[36] C.N. Anderson, P.P. Naulleau, P. Denham, D. Kemp, S. Rekawa, Dual-domain scanning illuminator for the SEMATECH Berkeley microfield exposure tool, *J. Vac. Sci. Technol. B Microelectron. Nanometer. Struct.* **25** (6) (2007) 2151–2154.
[37] G.F. Lorusso, A.M. Goethals, R. Jonckheere, J. Hermans, K. Ronse, A.M. Myers, et al., Extreme ultraviolet lithography at IMEC: Shadowing compensation and flare mitigation strategy, *J. Vac. Sci. Technol. B Microelectron. Nanometer. Struct.* **25** (6) (2007) 2127–2131.
[38] T. Abe, T. Adachi, H. Akizuki, H. Mohri, N. Hayashi, K. Ishikiriyama, EUV mask process development status for full field EUV exposure tool, in: *Proceedings of SPIE – The International Society for Optical Engineering*, SPIE, Bellingham, WA, USA, vol. 7028, 2008.
[39] N. Farrar, D. Brandt, J. Bonafede, Key parameters demonstrated for high-volume EUV lithography sources, *Semicond. Int.* **31** (2) (2008) 31–36.
[40] J. Mulkens, J. De Klerk, M. Leenders, F. De Jong, J.W. Cromwyk, Latest developments on immersion exposure systems. in: *Proceedings of SPIE – The International Society for Optical Engineering*, SPIE, Bellingham, WA, USA, vol. 6924, 2008.
[41] K. Tawarayama, S. Magoshi, Y. Tanaka, S. Shirai, H. Tanaka, Extreme ultraviolet lithography using small-field exposure tool: Current status, *Jpn. J. Appl. Phys., Part 1* **47** (6 PART 2) (2008) 4866–4871.
[42] S.Y. Chou, P.R. Krauss, P.J. Renstrom, Imprint of sub-25 nm vias and trenches in polymers, *Appl. Phys. Lett.* **67** (1995) 3114.
[43] S.Y. Chou, P.R. Krauss, W. Zhang, L. Guo, L. Zhuang, Sub-10 nm imprint lithography and applications, *J. Vac. Sci. Technol. B Microelectron. Nanometer. Struct.* **15** (6) (1997) 2897–2904.
[44] S.Y. Chou, C. Keimel, J. Gu, Ultrafast and direct imprint of nanostructures in silicon, *Nature* **417** (6891) (2002) 835–837.
[45] K.J. Morton, G. Nieberg, S. Bai, S.Y. Chou, Wafer-scale patterning of sub-40 nm diameter and high aspect ratio (>50:1) silicon pillar arrays by nanoimprint and etching, *Nanotechnology* **19** (34) (2008) 345303. doi:10.1088/0957-4484/19/34/34530.
[46] J. Choi, S.V. Sreenivasan, D. Resnick, UV nano-imprint lithography for manufacturing applications, in: *2007 Proceedings of the ASME International Design Engineering*

Technical Conferences and Computers and Information in Engineering Conference, IDETC2007 September 4–7, Novada, USA, vol. 3 PART B, 2008, pp. 985–991.

[47] W.H. Heath, F. Palmieri, J.R. Adams, B.K. Long, J. Chute, T.W. Holcombe, et al., Degradable cross-linkers and strippable imaging materials for step-and-flash imprint lithography, *Macromolecules* **41** (3) (2008) 719–726.

[48] Z.Z. Bandic, E.A. Dobisz, T.-W. Wu, T.R. Albrecht, Patterned magnetic media: Impact of nanoscale patterning on hard disk drives, *Solid State Technol.* **49** (Suppl. 9) (2006) S7–S13+S19.

[49] J. Wong, A. Scherer, M. Barbic, S. Schultz, Perpendicular patterned media in an $(Al_{0.9}Ga_{0.1})_2O_3$/GaAs substrate for magnetic storage, *J. Vac. Sci. Technol. B* **17** (8) (1999) 3190–3198.

[50] M.A.M. Haast, J.R. Schuurhuis, L. Abelmann, J.C. Lodder, Th.J. Popma, Reversal mechanism of submicron patterned CoNi/Pt multilayers, *IEEE J. Magn.* **34** (4) (1998) 1006–1008.

[51] A. Currie Munce, J.-U. Thiele, Hitachi's overseas research on hard disk drive, *Hitachi Rev.* **55** (2006) 150–154.

[52] E.A. Dobisz, Z.Z. Bandic, T.-W. Wu, T. Albrecht, Patterned media: Nanofabrication challenges of future disk drives, *Proc. IEEE* 96 (**11**) (2008) 1836–1846.

[53] J. Moritz, S. Landis, J.C. Toussaint, R. Bayle-Guillemaud, B. Rodmacq, G. Casali, et al., Patterned media made from pre-etched wafers: a promising route toward ultrahigh-density magnetic recording, *IEEE J. Magn.* **38** (4) (2002) 1731–1736.

[54] B.D. Terris, T. Thomson, Nanofabricated and self-assembled magnetic structures as data storage media, *J. Phys. D Appl. Phys.* **38** (12) (2005) R199–R222.

[55] M. Lutwyche, C. Andreoli, G. Binnig, J. Brugger, U. Drechsler, W. Häberle, et al., 5×5 2D AFM cantilever arrays a first step towards a terabit storage device, *Sens. Actuators A Phys.* **73** (1–2) (1999) 89–94.

[56] P. Vettiger, G. Binnig, The nanodrive project. Inventing a nanotechnology device for mass production and consumer use is trickier than it sounds, *Sci. Am.* **288** (1) (2003) 46–53.

[57] B.D. Terris, T. Thomas, G. Hu, Patterned media for future magnetic data storage, *Microsyst. Technol.* **13** (2007) 189–196.

[58] M.D. Austin, W. Zhang, H. Ge, D. Wasserman, S.A. Lyon, S.Y. Chou, 6 nm half-pitch lines and 0.04 μm^2 static random access memory patterns by nanoimprint lithography, *Nanotechnology* **16** (2005) 1058–1061.

[59] Y. Takagaki, E. Wiebicke, H. Kostial, K.H. Ploog, Fabrication of GHz-range surface-acoustic-wave transducers on $LiNbO_3$ using imprint lithography, *Nanotechnology* **13** (2002) 15–17.

[60] Y. Zhang, P. Kruit, Design of a high brightness multi-electron-beam source, in: *Physics Procedia*, vol. **1**, 2008, pp. 553–563.

[61] P. Kruit, The role of MEMS in maskless lithography, *Microelectron. Eng.* **84** (5–8) (2007) 1027–1032.

[62] J.R. Maldonado, Y. Sun, Z. Liu, X. Liu, S. Tanimoto, P. Pianetta, et al., Evaluation of electron energy spread in CsBr based photocathodes, *J. Vac. Sci. Technol. B Microelectron. Nanometer. Struct.* **26** (6) (2008) 2085–2090.

[63] M.L. Yu, S.T. Coyle, W. Devore, B. Shamoun, Electron–electron interaction induced beam displacement in a multiple electron beam system, *J. Vac. Sci. Technol. B Microelectron. Nanometer. Struct.* **23** (6) (2005) 2589–2595.

[64] M. Mankos, S. Coyle, A. Fernandez, A. Sagle, W. Owens, J. Sullivan, et al., Basic constraints for a multibeam lithography column, *J. Vac. Sci. Technol. B Microelectron. Nanometer. Struct.* **19** (2) (2001) 467–475.

[65] T.R. Groves,D.S. Pickard, Proposal for a distributed parallel system for high throughput maskless e-beam lithography, in: *Proceedings of SPIE – The International Society for Optical Engineering*, SPIE, Bellingham, USA, vol. 5504, 2004, pp. 173–177.

[66] D.S. Pickard, C. Kenney, S. Tanimoto, T. Crane, T. Groves, R.F.W. Pease, Monolithic multichannel secondary electron detector for distributed axis electron beam lithography and inspection, *J. Vac. Sci. Technol. B Microelectron. Nanometer. Struct.* **25** (6) (2007) 2277–2283.

[67] Q. Ji, X. Jiang, L. Ji, Y. Chen, B.V.D. Akker, K.-N. Leung, Novel ion beam tools for nanofabrication, in: *2005 NSTI Nanotechnology Conference and Trade Show – NSTI Nanotech 2005 Technical Proceedings*, 2005, pp. 703–706.

[68] D.S. Pickard, T.R. Groves, W.D. Meisburger, T. Crane, R.F. Pease, Distributed axis electron beam technology for maskless lithography and defect inspection, *J. Vac. Sci. Technol. B Microelectron. Nanometer. Struct.* **21** (6) (2003) 2834–2838.

[69] A.K. Sinensky, A.M. Belcher, Label-free and high-resolution protein/DNA nanoarray analysis using Kelvin probe force microscopy, *Nat. Nanotechnol.* **2** (10) (2007) 653–659.

[70] S.-W. Lee, C. Mao, C.E. Flynn, A.M. Belcher, Ordering of quantum dots, using genetically engineered viruses, *Science* **296** (5569) (2002) 892–895.

[71] Y. Huang, C.Y. Chiang, S.K. Lee, Y. Gao, E.L. Hu, J.D. Yoreo, et al., Programmable assembly of nanoarchitectures using genetically engineered viruses, *Nano Lett.* **7** (2005) 1429–1434.

[72] H.A. Becerril, P. Ludtke, B.M. Willardson, A.T. Woolley, DNA-templated nickel nanostructures and protein assemblies, *Langmuir* **22** (24) (2006) 10140–10144.

[73] X. Wang, F. Liu, G.T.S. Andavan, X. Jing, K. Singh, V.R. Yazdanpanah, et al., Carbon nanotube-DNA nanoarchitectures and electronic functionality, *Small* **2** (11) (2006) 1356–1365.

[74] S. Huang, E. Schopf, Y. Chen, Dynamically configurable biomolecular nanoarrays, *Nano Lett.* **7** (10) (2007) 3116–3121.

[75] L. Iversen, N. Cherouati, T. Berthing, D. Stamou, K.L. Martinez, Templated protein assembly on micro-contact-printed surface patterns. Use of the snap-tag protein functionality, *Langmuir* **24** (12) (2008) 6375–6381.

[76] D.A. Dehlinger, B.D. Sullivan, S. Esener, M.J. Heller, Electric-field-directed assembly of biomolecular-derivatized nanoparticles into higher-order structures, *Small* **3** (7) (2007) 1237–1244.

[77] R. Chhabra, J. Sharma, Y. Ke, Y. Liu, S. Rinker, S. Lindsay, et al., Spatially addressable multiprotein nanoarrays templated by aptamer-tagged DNA nanoarchitectures, *J. Am. Chem. Soc.* **129** (34) (2007) 10304–10305.

[78] S. Rinker, Y. Ke, Y. Liu, R. Chhabra, H. Yan, Self-assembled DNA nanostructures for distance-dependent multivalent ligand-protein binding, *Nat. Nanotechnol.* **3** (7) (2008) 418–422.

[79] A.R. Hirst, B. Escuder, J.F. Miravet, D.K. Smith, High-tech applications of self-assembling supramolecular nanostructured gel-phase materials: from regenerative medicine to electronic devices, *Angew. Chem. Int. Ed. Eng.* **47** (42) (2008) 8002–8018.

[80] S. Chaterji, I.K. Kwon, K. Park, Smart polymeric gels: Redefining the limits of biomedical devices, *Prog. Polym. Sci.* **32** (8–9) (2007) 1083–1122.

[81] M.-L. Visnapuu, T. Fazio, S. Wind, E.C. Greene, Parallel arrays of geometric nanowells for assembling curtains of DNA with controlled lateral dispersion, *Langmuir* **24** (19) (2008) 11293–11299.

[82] J. Tang, A. Ebner, H. Badelt-Lichtblau, C. Völlenkle, C. Rankl, B. Kraxberger, et al., Recognition imaging and highly ordered molecular templating of bacterial S-layer nanoarrays containing affinity-tags, *Nano Lett.* 2008 **8** (12), pp. 4312–4319.

[83] G.-J. Zhang, T. Tanii, Y. Kanari, I. Ohdomari, Production of nanopatterns by a combination of electron beam lithography and a self-assembled monolayer for an antibody nanoarray, *J. Nanosci. Nanotechnol.* **7** (2) (2007) 410–417.

[84] H. Chen, J. Li, Nanotechnology: Moving from microarrays toward nanoarrays, *Methods Mol. Biol.* **381** (2007) 411–436.

[85] A. Montsant, U. Maheswari, C. Bowler, P.J. Lopez, Diatomics: Toward diatom functional genomics, *J. Nanosci. Nanotechnol.* **5** (1) (2005) 5–14.

[86] M. Lynch, C. Mosher, J. Huff, S. Nettikadan, J. Johnson, E. Henderson, Functional protein nanoarrays for biomarker profiling, *Proteomics* **4** (6) (2004) 1695–1702.

[87] D.B. Allred, A. Cheng, M. Sarikaya, F. Baneyx, D.T. Schwartz, Three-dimensional architecture of inorganic nanoarrays electrodeposited through a surface-layer protein mask, *Nano Lett.* **8** (5) (2008) 1434–1438.

[88] J. Henzie, J. Lee, M.H. Lee, W. Hasan, T.W. Odom, Nanofabrication of plasmonic structures, *Annu. Rev. Phys. Chem.* **60** (2009) 127–146.

[89] G. Duan, W. Cai, Y. Luo, Z. Li, Y. Lei, Hierarchical structured Ni nanoring and hollow sphere arrays by morphology inheritance based on ordered through-pore template and electrodeposition, *J. Phys. Chem. B* **110** (32) (2006) 15729–15733.

[90] H.-J. Her, J.-M. Kim, J. Kim, Y.J. Choi, C.J. Kang, Y.-S. Kim, Fabrication of titania nanopoles by nanoporous alumina template, in: *Proceedings of the 2nd IEEE International Conference on Nano/Micro Engineered and Molecular Systems*, NEMS '07, 2007, IEEE Service Center, Piscataway, NJ, USA, pp. 788–791.

[91] D.H. Kim, S.Y. Park, Y.D. Park, J.H. Kim, K. Sun, G.H. Lee, et al., Cellular responses to nanotopology of polymeric surfaces fabricated with AAO nanoimprinting, in: *Proceedings of the 3rd IEEE/EMBS Special Topic Conference on Microtechnology in Medicine and Biology*, 2005, Piscataway, NJ, USA, pp. 188–189.

[92] E. Lamers, X. Frank Walboomers, M. Domanski, J. te Riet, F.C.M.J.M. van Delft, R. Luttge, et al., The influence of nanoscale grooved substrates on osteoblast behavior and extracellular matrix deposition, *Biomaterials* **31** (12) (2010) 3307–3316.

[93] E.I. Meletis, J.C. Jiang, Ordered, self-organized cobalt nanodots in Co-diamond-like carbon thin films, *J. Nanosci. Nanotechnol.* **6** (6) (2006) 1807–1810.

[94] A. Hozumi, S. Asakura, A. Fuwa, N. Shirahata, Photochemical fabrication of a well-defined diblock copolymer nanotemplate using 172-nm vacuum ultraviolet light, *J. Colloid Interface Sci.* **285** (2) (2005) 875–878.

[95] B. Kim, M.A. Carignano, S.L. Tripp, A. Wei, Cluster size analysis of two-dimensional order in colloidal gold nanoparticle arrays, *Langmuir* **20** (2004) 9360–9365.

[96] C. Tang, E.M. Lennon, G.H. Fredrickson, E.J. Kramer, C.J. Hawker, Evolution of block copolymer lithography to highly ordered square arrays, *Science* **322** (5900) (2008) 429–432.

[97] X. Zhang, E.M. Hicks, J. Zhao, G.C. Schatz, R.P. Van Duyne, Electrochemical tuning of silver nanoparticles fabricated by nanosphere lithography, *Nano Lett.* **5** (7) (2005) 1503–1507.

[98] I. Bita, J.K.W. Yang, Y.S. Jung, C.A. Ross, E.L. Thomas, K.K. Berggren, Graphoepitaxy of self-assembled block copolymers on two-dimensional periodic patterned templates, *Science* **321** (5891) (2008) 939–943.

[99] K.Ch. Daoulas, M. Müller, M.P. Stoykovich, H. Kang, J.J. de Pablo, P.F. Nealey, Directed copolymer assembly on chemical substrate patterns: A phenomenological and single-chain-in-mean-field simulations study of the influence of roughness in the substrate pattern, *Langmuir* **24** (4) (2008) 1284–1295.

[100] C.T. Black, Block copolymers: Nanowire arrays build themselves, *Nat. Nanotechnol.* **2** (8) (2007) 464–465.

[101] X.-F. Wu, Y.A. Dzenis, Guided self-assembly of diblock copolymer thin films on chemically patterned substrates, *J. Chem. Phys.* **125** (17) (2006) 174707.

[102] J.Y. Cheng, A.M. Mayes, C.A. Ross, Nanostructure engineering by templated self-assembly of block copolymers, *Nat. Mater.* **3** (11) (2004) 823–828.

[103] R. Glass, M. Möller, J.P. Spatz, Block copolymer micelle nanolithography, *Nanotechnology* **14** (2003) 1153/1160.

[104] B. Platschek, N. Petkov, D. Himsl, S. Zimdars, Z. Li, R. Köhn, et al., Vertical columnar block-copolymer-templated mesoporous silica via confined phase transformation, *J. Am. Chem. Soc.* **130** (51) (2008) 17362–17371.

[105] R. Ruiz, H. Kang, F.A. Detcheverry, E. Dobisz, D.S. Kercher, T.R. Albrecht, et al., Density multiplication and improved lithography by directed block copolymer assembly, *Science* **321** (5891) (2008) 936–939.

[106] W. Fan, M.A. Snyder, S. Kumar, P.-S. Lee, W. C. Yoo, A.V. McCormick, et al., Hierarchical nanofabrication of microporous crystals with ordered mesoporosity, *Nat. Mater.* **7** (12) (2008) 984–991.

[107] M.L. Schattenburg, R.J. Aucoin, R.C. Fleming, Optically matched trilevel resist process for nanostructure fabrication, *J. Vac. Sci. Technol. B* **13** (6) (1995) 3007–3011.

[108] X. Chen, S.H. Zaidi, S.R.J. Brueck, D.J. Devinec, Interferometric lithography of sub-micrometer sparse hole arrays for field-emission display applications, *J. Vac. Sci. Technol. B* **14** (5) (1996) 3339–3349.

[109] S.R.J. Brueck, Optical and interferometric lithography nanotechnology enablers, *Proc. IEEE 93* (2005) 1704–1721.

[110] D. López, V.A. Aksyuk, G.P. Watson, W.M. Mansfield, R. Cirelli, F. Klemens, et al., Two-dimensional MEMS array for maskless lithography and wavefront modulation,

in: Proceedings of SPIE – The International Society for Optical Engineering, SPIE, Bellingham, WA, USA, vol. 6589, 2007.

[111] G.P. Watson, V. Aksyuk, M.E. Simon, D.M. Tennant, R.A. Cirelli, W.M. Mansfield, et al., Spatial light modulator for maskless optical projection lithography, *J. Vac. Sci. Technol. B Microelectron. Nanometer. Struct.* **24** (6) (2006) 2852–2856.

[112] S. Pau, G.P. Watson, O. Nalamasu, Writing an arbitrary non-periodic pattern using interference lithography, *J. Mod. Opt.* **48** (7) (2001) 1211–1223.

[113] W.K. Choi, T.H. Liew, M.K. Dawood, H.I. Smith, C.V. Thompson, M.H. Hong, Synthesis of silicon nanowires and nanofin arrays using interference lithography and catalytic etching, *Nano Lett.* **8** (11) (2008) 3799–3802.

[114] H.H. Solak, Nanolithography with coherent extreme ultraviolet light, *J. Phys. D Appl. Phys.* **39** (2006) R171–R188.

[115] N. Choksi, D.S. Pickard, M. McCord, R.F.W. Pease, Y. Shroff, Y. Chen, et al., Maskless extreme ultraviolet lithography, *J. Vac. Sci. Technol. B Microelectron. Nanometer. Struct.* **17** (6) (1999) 3047–3051.

[116] J.-S. Wang, S. Hafeman, A.R. Neureuther, O. Solgaard, Effects of through-focus symmetry in maskless lithography using micromirror arrays, *J. Vac. Sci. Technol. B Microelectron. Nanometer. Struct.* **23** (6) (2005) 2738–2742.

[117] I.W. Jung, J.-S. Wang, O. Solgaard, Spatial light modulators for maskless lithography, in: *IEEE/LEOS International Conference on Optical MEMS and Their Applications* 2006, Piscataway, NJ, USA, pp. 150–151.

[118] A. Isoyan, Y.-C. Cheng, F. Jiang, J. Wallace, F. Cerrina, S. Bollepalli, Progress in extreme ultraviolet interferometric and holographic lithography, *J. Vac. Sci. Technol. B Microelectron. Nanometer. Struct.* **25** (6) (2007) 2145–2150.

[119] J. Wallace, Y.-C. Cheng, A. Isoyan, Q. Leonard, M. Fisher, M. Green, et al., A novel EUV exposure station for nanotechnology studies, *Nucl. Instrum. Methods Phys. Res. A* **582** (1) (2007) 254–257.

[120] A. Isoyan, A. Wüest, J. Wallace, F. Jiang, F. Cerrina, 4X reduction extreme ultraviolet interferometric lithography, *Opt. Express* **16** (12) (2008) 9106–9111.

[121] J.J. McClelland, R. Gupta, R.J. Celotta, Laser focusing of chromium atoms for nanostructure fabrication, in: *Proceedings of the European Quantum Electronics Conference* EQEC '96, 8–13 September 1996, 1996, pp. 154–154.

[122] J.J. McClelland, S.B. Hill, M. Pichler, A fast, deterministic source of single Cr atoms, in: *Proceedings of the QELS Quantum Electronics and Laser Science Postconference Digest*, 1–6 June 2003, 2003, pp. 2pp.

[123] F. Perales, J. Robert, J. Baudon, M. Ducloy, Ultra thin coherent atom beam by Stern–Gerlach interferometry, *Europhys. Lett.* **78** (6) (2007) doi: 10.1209/0295-5075/78/60003.

[124] S. Kuiper, M. De Boer, C. Van Rijn, W. Nijdam, G. Krijnen, M. Elwenspoek, Wet and dry etching techniques for the release of sub-micrometre perforated membranes, *J. Micromech. Microeng.* **10** (2) (2000) 171–174.

[125] L. Vogelaar, W. Nijdam, H.A.G.M. van Wolferen, R.M. de Ridder, F.B. Segerink, E. Fluck, et al., Large area photonic crystal slabs for visible light with waveguiding

defect structures: Fabrication with focused ion beam assisted laser interference lithography, *Adv. Mater.* **13** (2001) 1551–1554.

[126] C.G. Bostan, R.M. de Ridder, I. van Dorssen, H.A.G.M. van Wolferen, L. Kuipers,N.F. van Hulst, Design for laser interference lithography used in fabricating two-dimensional periodic structures, in: *ICTON-2002 Conference Proceedings*, 2002, p. 61.

[127] R. Murillo, H.A. van Wolferen, L. Abelmann, J.C. Lodder, Fabrication of patterned magnetic nanodots by laser interference lithography, *Microelectron. Eng.* **78–79** (2005) 260–265.

[128] A.J. le Febre, R. Luttge, L. Abelmann, J.C. Lodder, Field emission to control tip-sample distance in magnetic probe recording, *J. Phys. Conf. Ser.* **61** (2007) 673–677.

[129] J.Y. Cheng, C.A. Ross, E.L. Thomas, H.I. Smith, G.J. Vancso, Fabrication of nanostructures with long-range order using block copolymer lithography, *Appl. Phys. Lett.* **81** (19) (2002) 3657.

[130] N.V. Dziomkina, M.A. Hempenius, G.J. Vancso, Symmetry control of polymer collodial monolayers and crystals by electrophoretic deposition onto patterned surfaces, *Adv. Mater.* **17** (2005) 237–240.

[131] D.S. Kim, R. Ji, H.J. Fan, F. Bertram, R. Scholz, A. Dadgar, et al., Laser-interference lithography tailored for highly symmetrically arranged ZnO nanowire arrays, *Small* **3** (1) (2007) 76–80.

[132] P. Maury, M. Escalante, D.N. Reinhoudt, J. Huskens, Directed assembly of nanoparticles onto polymer-imprinted or chemically patterned templates fabricated by nanoimprint lithography, *Adv. Mater.* **17** (22) (2005) 2718–2723.

[133] R. Luttge, H.A.G.M. Van Wolferen, L. Abelmann, Laser interferometric nanolithography using a new positive chemical amplified resist, *J. Vac. Sci. Technol. B Microelectron. Nanometer. Struct.* **25** (6) (2007) 2476–2480.

[134] M. Domanski, R. Luttge, E. Lamers, F. Walboomers, A.J.A. Winnubst, J.A. Jansen, et al., Large-area silicon masters for the fabrication of nanostructured bioactive surfaces utilizing laser interference nanolithography and dry etching, *EIPBN* 2009, Marco Island, Florida, May 26–29, 2009.

[135] M. Domanski, F. Lamers, E. Walboomers, A.J.A. Winnubst, J. Jansen, R. Luttge, et al., Nanosurf-nanostructured biomaterial surfaces, in: *Proceedings of the 12th Annual European Conference on Micro- and Nanoscale Technologies for the Biosciences*, 17–19 November, Montreux, Switzerland, 2008.

[136] L. Prodanov, J. te Riet, E. Lamers, M. Domanski, R. Luttge, J.J.W.A. van Loon, et al., The interaction between nanoscale surface features and mechanical loading and its effect on osteoblast-like cells behavior, *Biomaterials* **31** (30) (2010) 7758–7765.

[137] A.J. Torres, M. Wu, D. Holowka, B. Baird, Nanobiotechnology and cell biology: Micro- and nanofabricated surfaces to investigate receptor-mediated signaling, *Annu. Rev. Biophys.* **37** (2008) 265–288.

[138] G. Mendonca, D.B.S. Mendonca, F.J.L. Arago, L.F. Cooper, Advancing dental implant surface technology – from micron- to nanotopography, *Biomaterials* **29** (28) (2008) 3822–3835.

[139] F.C.M.J.M. van Delft, F.C. van den Heuvel, W.A. Loesberg, J. te Riet, P. Schön, C.G. Figdor, et al., Manufacturing substrate nano-grooves for studying cell alignment and adhesion, *Microelectron. Eng.* **85** (5–6) (2008) 1362–1366.

[140] J.A.T.C. Parker, X.F. Walboomers, J.W. Von Den Hoff, J.C. Maltha, J.A. Jansen, Soft-tissue response to silicone and poly-L-lactic acid implants with a periodic or random surface micropattern, *J. Biomed. Mater. Res.* **61** (1) (2002) 91–98.

[141] M. Domanski, R. Luttge, E. Lamers, A.J.A. Winnubst, F.X. Walboomers, J.A. Jansen, et al., Novel approach to produce nanopatterned titanium implants by combining nanoimprint lithography and reactive ion etching, in: *Proceedings of MicroTAS 2010, 14th International Conference on Miniaturized Systems in Chemistry and Life Sciences*, 3–7 October, Groningen, The Netherlands, 2010.

[142] S.W. Pang, Flexible 3D channels and cavities formation by nanoimprinting, in: *Proceedings of the 5th IEEE Conference on Nanotechnology*, vol. 2, 2005, pp. 450–451.

[143] D.C. Meisel, M. Wegener, K. Busch, Three-dimensional photonic crystals by holographic lithography using the umbrella configuration: Symmetries and complete photonic band gaps, *Phys. Rev. B* **70** (2004) 165104–165110.

[144] P. Lisboa, P. Gililand D. Valsesia, A. Colpo, G. Ceccone, A. Papadopoulou-Bouraoui, et al., Thiolated polyethylene oxide as a non-fouling element for nano-patterned bio-devices, *Appl. Surf. Sci.* **253** (10) (2007) 4796–4804.

[145] A. Valsesia, Fabrication of nanostructured surfaces for the development of advanced biointerfaces, *Sci. Acta* **1** (1) (2007) 153–157.

[146] F. Hui, B. Li, P. He, J. Hu, Y. Fang, Electrochemical fabrication of nanoporous polypyrrole film on HOPG using nanobubbles as templates, *Electrochem. Commun.* **11** (3) (2008) 639–642.

[147] P.M. George, A.W. Lyckman, D.A. Lavan, A. Hegde, Y. Leung, R. Avasare, et al., Fabrication and biocompatibility of polypyrrole implants suitable for neural prosthetics, *Biomaterials* **26** (17) (2005) 3511–3519.

[148] C.-M. Shih, S.-J. Lin, Y.-Y. Su, C.-C. Shih, Amorphous oxide – a platform for drug delivery, *J. Controlled Release* **102** (2005) 539–549.

[149] S.C. Mailley, M. Hyland, P. Mailley, J.M. McLaughlin, E.T. McAdams, Electrochemical and structural characterizations of electrodeposited iridium oxide thin-film electrodes applied to neurostimulating electrical signal, *Mater. Sci. Eng. C* **21** (1-2) (2002) 167–175.

[150] W.-D. Huang, J. Wang, T. Ativanichayaphong, M. Chiao, J.C. Chiao, Development of an IrOx micro pH sensor array on flexible polymer substrate, in: *Proceedings of SPIE – The International Society for Optical Engineering*, vol. 6931, 2008.

[151] P. Yeh, C.S. Yariv, A. Hong, Electromagnetic propagation in periodic stratified media. 1. General theory, *J. Opt. Soc. Am.* **67** (1977) 423–438.

[152] R.J. Gehr, R.W. Boyd, Optical properties of nanostructured optical materials, *Chem. Mater.* **8** (1996) 1807–1819.

[153] E. Cazzanelli, M. Castriota, S. Marino, N. Scaramuzza, J. Purans, A. Kuzmin, et al., Characterization of rhenium oxide films and their application to liquid crystal cells, *Appl. Phys. Lett.* **105** (2009) 114904.

[154] K. Wang, A. Reznik, K.S. Karim, Fabrication and characterization of cerium oxide thin films for ultraviolet sensing applications, in: *Proceedings of SPIE – The International Society for Optical Engineering*, vol. 7425, 2009.

[155] S. Saha, S.K. Arya, S.P. Singh, K. Sreenivas, B.D. Malhotra, V. Gupta, Nanoporous cerium oxide thin film for glucose biosensor, *Biosens. Bioelectron.* **24** (7) (2009) 2040–2045.

[156] S.G. Ansari, Z.A. Ansari, H.-K. Seo, G.-S. Kim, Y.-S. Kim, G. Khang, et al., Urea sensor based on tin oxide thin films prepared by modified plasma enhanced CVD, *Sens. Actuators B Chem.* **132** (1) (2008) 265–271.

[157] C. Rhodes, S. Franzen, J.-P. Maria, M. Losego, D.N. Leonard, B. Laughlin, Surface plasmon resonance in conducting metal oxides, *Appl. Phys. Lett.*, **100** (5) (2006).

[158] L. Vayssieres, On the design of advanced metal oxide nanomaterials, *Int. J. Nanotechnol.* **1** (1–2) (2004) 1–41.

[159] U. Schubert, S. Tewinkel, R. Lamber, Metal complexes in inorganic matrixes. 15.1 coordination of metal ions by lysinate-modified titanium and zirconium alkoxides and the preparation of metal/titania and metal/zirconia nanocomposites, *Chem. Mater.* **8** (8) (1996) 2047–2055.

[160] J. Ravichandran, W. Siemons, M.L. Scullin, S. Mukerjee, M. Huijben, J.E. Moore, et al., Understanding the nature of electronic effective mass in double-doped $SrTiO_3$. http://arxiv.org/PS_cache/arxiv/pdf/0809/0809.4706v2, (arXiv:0809.4706v2), 2009.

[161] S. Beke, L. Korösi, S. Papp, L. Nánai, A. Oszkó, J.G. Kiss, et al., Nd:YAG laser synthesis of nanostructural V2O5 from vanadium oxide sols: Morphological and structural characterizations, *Appl. Surf. Sci.* **254** (5) (2007) 1363–1368.

[162] S. Beke, S. Giorgio, L. Korösi, L. Nánai, W. Marine, Structural and optical properties of pulsed laser deposited V2O5 thin films, *Thin Solid Films* **516** (15) (2008) 4659–4664.

[163] Y. Cai,B.M. Ocko, Electro pen nanolithography, *J. Am. Chem. Soc.* **127** (46) (2005) 16287–16291.

[164] Y. Li, B.W. Maynor, J. Liu, Electrochemical AFM dip-pen nanolithography, *J. Am. Chem. Soc.* **123** (9) (2001) 2105–2106.

[165] R. Garcia, R.V. Martinez, J. Martinez, Nano-chemistry and scanning probe nanolithographies, *Chem. Soc. Rev.* **35** (2006) 29–38.

[166] E. Reimhult, K. Kumar, W. Knoll, Fabrication of nanoporous silicon nitride and silicon oxide films of controlled size and porosity for combined electrochemical and waveguide measurements, *Nanotechnology* **18** (27) (2007) 275303.

[167] P.T. Lin, Z. Zhifu Liu, B.W. Wessels, Simulation and fabrication of two dimensional nonlinear photonic crystals using barium titanate thin films. *MRS Proceedings*, 2007.

[168] J.H. Lee, K.Y. Song, H.J. Yoon, J.S. Kim, T. Hasegawa, T. Nagashima, et al., Brillouin gain-coefficient measurement for bismuth-oxide-based photonic crystal fiber under significant beam reflection at splicing points, *Opt. Lett.* **34** (2009) 2670–2671.

[169] M.R. De Guire, T.P. Niesen, S. Supothina, J. Wolff, J. Bill, C.N. Sukenik, et al., Synthesis of oxide and non-oxide inorganic materials at organic surfaces, *Zeitschrift fuer Metallkunde/Mater. Res. Adv. Tech.* **89** (11) (1998) 758–766.

[170] A.A. Yasseri, S. Sharma, T.I. Kaminsc, Q. Xia, S.Y. Chou, R.F.W. Pease, Alkylsiloxane self-assembly monolayer formation guided by nanoimprinted Si and SiO_2 templates, *Appl. Phys. Lett.* **89** (2006) 153121.

[171] S. Friebel, J. Aizenberg, S. Abad, P. Wiltzius, Ultraviolet lithography of self-assembled monolayers for submicron patterned deposition, *Appl. Phys. Lett.* **77** (5) (2000) 2406–2408.

[172] N.F. Steinmetz, S.N. Shah, J.E. Barclay, G. Rallapalli, G.P. Lomonossof, D.J. Evans, Virus-templated silica nanoparticles, *Small* **5** (7) (2009) 813–816.

[173] J. Han, H. Su, F. Song, J. Gu, D. Zhang, L. Jiang, Novel photonic crystals: Incorporation of nano-CdS into the natural photonic crystals within peacock feathers, *Langmuir* **25** (5) (2009) 3207–3211.

[174] H. Su, Q. Dong, J. Han, D. Zhang, Q. Guo, Biogenic synthesis and photocatalysis of Pd-PdO nanoclusters reinforced hierachical TiO_2 films with interwoven and tubular conformations, *Biomacromolecules* **9** (2) (2008) 499–504.

[175] W. Zhang, D. Zhang, T. Fan, J. Gu, J. Ding, H. Wang, et al., Novel photoanode structure templated from butterfly wing scales, *Chem. Mater.* **21** (1) (2009) 33–40.

[176] A.R. Parker, H.E. Townley, Biomimetic of photonic nanostructures, *Nat. Nanotechnol.* **2** (6) (2007) 347–353.

[177] A.R. Parker, Natural photonics for industrial inspiration, *Philos. Trans. R. Soc. A* **367** (2009) 1759–1782.

[178] T. Lohmüller, M. Helgert, M. Sundermann, R. Brunner, J.P. Spatz, Biomimetic interfaces for high-performance optics in the deep-UV light range, *Nano Lett.* **8** (5) (2008) 1429–1433.

[179] S. Balci, A.M. Bittner, M. Schirra, K. Thonke, R. Sauer, K. Hahn, et al., Catalytics coating of virus particles with zinc oxide, *Electrochim. Acta* **54** (22) (2009) 5149–5154.

[180] J. Liu, Y. Mao, E. Lan, D.R. Banatao, G.J. Forse, J. Lu, et al., Generation of oxide nanopatterns by combining self-assembly of S-layer proteins and area-selective atomic layer deposition, *J. Am. Chem. Soc.* **130** (50) (2008) 16908–16913.

[181] M. Sarikaya, C. Tamerler, A.K.-Y. Jen, K. Schulten, F. Baneyx, Molecular biomimetics: Nanotechnology through biology, *Nat. Mater.* **2** (9) (2003) 577–585.

[182] H. Gao, J. Henzie, T.W. Odom, Direct evidence for surface plasmon-mediated enhanced light tansmission through metallic nanohole arrays, *Nano Lett.* **6** (9) (2006) 2104–2108.

[183] K.-J. Byeon, S.-Y. Hwang, H. Lee, Fabrication of nano-hole array patterns on transparent conducting oxide layer using thermally curable nanoimprint lithography, *Microelectron. Eng.* **85** (5–6) (2008) 830–833.

[184] M. Castriota, T. Caruso, A. Policicchio, S. La Rosa, R.G. Agostino, E. Cazzanelli, Anomalous enhancement of raman scattering of metal oxide film deposited on thermally treated ITO-coated glass substrates, *Chem. Phys. Lett.* **478** (4–6) (2009) 195–199.

[185] H. Chen, J. Li, Nanotechnology: Moving from microarrays toward nanoarrays, *Methods Mol. Biol.* **381** (2007) 411–436.

[186] L.L. Chan, M. Pineda, J.T. Heeres, P.J. Hergenrother, B.T. Cunningham, A general method for discovering inhibitors of protein-DNA interactions using photonic crystal biosensors, *ACS Chem. Biol.* **3** (7) (2008) 437–448.

[187] A. Valsesia, I. Mannelli, P. Colpo, F. Bretagnol, F. Rossi, Protein nanopatterns for improved immunodetection sensitivity, *Anal. Chem.* **80** (19) (2008) 7336–7340.

[188] J.L. West, N.J. Halas, Engineered nanomaterials for biophotonics applications: Improving sensing, imaging, and therapeutics, *Annu. Rev. Biomed. Eng.* **5** (2003) 285–292.

[189] H. Perry, A. Gopinath, D.L. Kaplan, L.D. Negro, F.G. Omenetto, Nano- and micropatterning of optically transparent, mechanically robust, biocompatible silk fibroin films, *Adv. Mater.* **20** (16) (2008) 3070–3072.

[190] T. Hamaguchi, M. Uno, S. Yamanaka, Photocatalytic activity of titania nanohole arrays, *J. Photochem. Photobiol. A* **173** (1) (2005) 99–105.

[191] Y. Li, T. Sasaki, Y. Shimizu, N. Koshizaki, Hexagonal-close-packed, hierarchical amorphous TiO_2 nanocolumn arrays: Transferability, enhanced photocatalytic activity, and superamphiphilicity without UV irradiation, *J. Am. Chem. Soc.* **130** (44) (2008) 14755–14762.

[192] S.D. Caruthers, S.A. Wickline, G.M. Lanza, Nanotechnological applications in medicine, *Curr. Opin. Biotechnol.* **18** (2007) 26–30.

[193] U. Feldkamp, C.M. Niemeyer, Rational design of DNA nanoarchitectures, *Angew. Chem. Int. Ed. Engl.* (2006) 1856–1876.

CHAPTER

Micromechanical Transducers

5

CHAPTER CONTENTS

Miniaturized micromechanical-based transducers are applied in many various scientific communities and industries. In this chapter we look into the basic concepts of this extensive and important category of devices using simple components as examples. Transducers come in such extreme diversity that a descriptive overview such as is provided by this book is best taken from the perspective of the user. One of the most easily understood definitions of transducers is that they convert energy from one form into another. This perspective also shows that there is no one single design rule. The designer needs to start from a range of actual conditions to realize the required transduction principle in a given application.

Microfabrication for Industrial Applications. DOI: 10.1016/B978-0-8155-1582-1.00005-8

At its heart a transducer is a component of a measurement system; the transducers at the input and the output of a measurement system are called sensors and actuators, respectively. We distinguish six different energy domains that need to be measured: radiant, mechanical, thermal, electrical, magnetic and chemical [1]. Transducers are present in both technical and biological systems. A great many transduction processes occur in natural (biological) systems, which are widely studied, mainly with respect to their importance in understanding disease and possible therapeutic interventions using pharmaceuticals. Researchers from all disciplines have investigated the functions appearing in natural systems for many decades, and recently more and more engineered systems have demonstrated that these natural functions can successfully inspire novel sensor and actuator designs.

Initial industrial developments focused on transduction in technically dominated environments, but there are many examples where technical and biological systems meet (medicine, biopharmaceutics, agriculture and the bioindustry). This chapter introduces the field of micromechanical sensors and actuators and their technical applications. In this context, the mechanical sensors and actuators that convert mechanical energy into electrical energy and vice versa are particularly interesting, since these devices are so essential to the design of many technical systems. Figure 5.1 presents a simplistic flow-chart of a system at a very abstract level.

The following sections will cover a variety of material and design aspects that are relevant to micro- and nanomechanical transducers. Any of the transducers which are introduced here as microactuators may be used in reversed operation as a sensor, and a vast amount of literature exists that describes micromechanical transducers in sensor applications. This field is too extensive to discuss details of engineering design and the theoretical background of sensors and actuators within the context of this book, hence we will focus on selected microfabrication technologies and their applications. The critical issues of packaging microfabricated sensors for real-world applications, however, are briefly mentioned.

I realize that the choice of topics covered here may appear to be randomly selected. To answer this, I would point out that the importance of this field lies not with any single new phenomenon, but the fact that our lives have been changed by developments from different disciplines and their integration into one system. This means that it does not much matter which examples are selected from the many currently being investigated. It is, however, important to mention the fairly novel combinations, such as optomechanical,

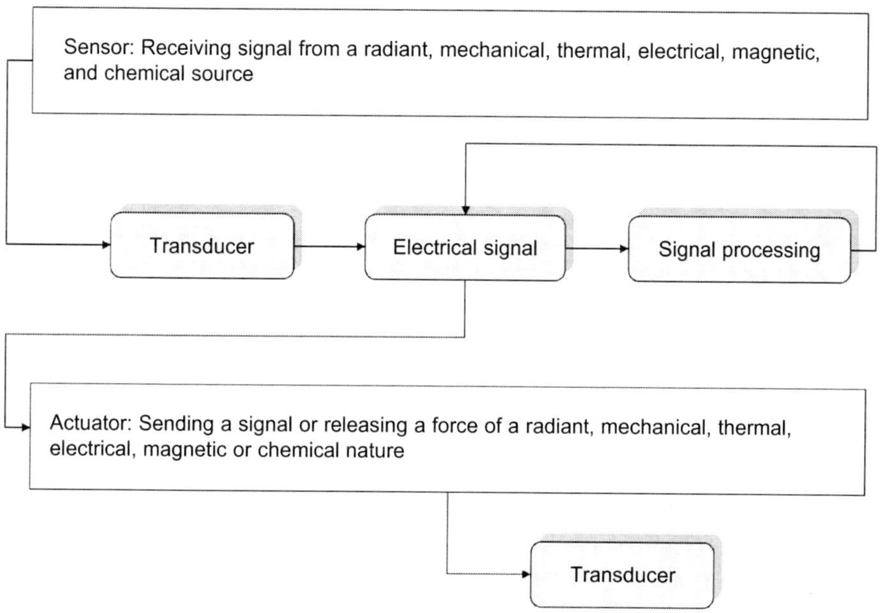

FIGURE 5.1

Very simplified technical interpretation of a system. This overview illustrates the central role of transducers.

electrochemical-mechanical, and (bio)chemomechanical devices, that have arisen in this field. A wealth of further reading is available on the details of microscale transducers and their backgrounds [2–5]. The examples presented here may be taken as an initial starting point for study in this field of research and development. Students may use the following sections as a guideline to the problems and design aspects of the various sensing and actuation principles, e.g., factors such as sensitivity, temperature dependency or material strength.

In the following section we will briefly introduce the field of applications, and after this we will discuss materials and methods in general and silicon resonant devices in particular. In this chapter, I will also discuss a hands-on case for the application of a resonant actuator based on my PhD project. This relatively simple example could be used in the practical part of a microfabrication course. The topic of modeling of such systems is beyond the scope of this book, and readers are referred to the examples published elsewhere [6].

5.1 APPLICATION FIELDS

When utilizing resonating elements as a sensor, a system parameter (e.g., mass, stiffness, strain) is changed. A shift of the eigenvalue of the system will then occur and this produces a readout for the measurand in the frequency domain. The miniaturization of such transducers is driven by a need for light weight, low cost of manufacture and ease of portability by hand-held formats. Ideally, these needs are fulfilled by the utilization of microsystems rather than precision engineering and fine-mechanical assembly. The latter may be possible but it is cumbersome and faces challenging structural limitations around lateral dimensions of 100 μm.

Although it is possible to build dedicated precision engineering assembly lines for structures in the micrometer range, many of the microfabricated sensors and actuator devices have actually been first demonstrated by utilizing the silicon micromachining technologies discussed in Chapters 2 and 3. These silicon sensor devices are more sensitive than their precision engineered counterparts. In the light of current economic demands, polymer electronics, and re-engineering of proven silicon microsystems into a polymer replication process can be considered as a possibility for large-scale production. A paradigm shift in materials selection has occurred because nanoscale sensors and actuators often rely on self-assembly at the scale of molecules and atoms, which enormously broadens the diversity of potentially applicable materials.

Design modeling is very important in the field of micromechanical sensors and actuators. A number of analytical models have been investigated and compared to numerical models of device designs. Understanding the governing formulae for theoretical and experimental modeling is a prerequisite for the optimization of such integrated systems. These sophisticated design studies for the fine-tuning of structural and energetic performance are important, because correction of the devices is often not possible once they reach the back-end of the process. Re-running a complete process would be extremely expensive. In many research projects, access to a microfabrication infrastructure often leads to the creation of demonstrators without any model-based design iterations. However, in commercial production environments modeling is extensively used to avoid design mistakes, and therefore costs.

As yet, only a few microelectromechanical devices or components have been successfully introduced into the real world. Reduction of design complexity and greater levels of accuracy need to be achieved first. The practical hands-on approach of research can help, because actually building the device and observing is still far easier than modeling it from scratch. It is important to realize that the modeling tools existing in industry are often based on the aforementioned observations and discoveries at the demonstrator level. While a reduction of complexity will improve production yields per batch, only better performance will make the devices competitive.

A type of micromachined sensor that is used in a multitude of industrial applications is the pressure sensor. Figure 5.2 (left) depicts the *SMD085*, which is one of the first micromechanical pressure sensors in a standardized SMD (surface mounted device) package, manufactured by Bosch. The sensor measures barometric altitude for an ambient pressure range of 0.6 to 1.15 bar. Figure 5.2 (right) depicts the new digital MEMS pressure sensor *BMP085* of Bosch Sensortec, shown in its device package of $5 \times 5 \times 1.2$ millimeters. The sensor is fabricated to conform with current developments and is intended to integrate navigation functions into mobile phones. The sensor has a resolution of up to 0.25 meters in altitude change and shows full conformity to the Restriction of Hazardous Substances (RoHS) directives, and is also halogen-free.

The fabrication of such a system undergoes stringent process control conditions in Bosch's production environment at Reutlingen, Germany. Figure 5.3 depicts such a process example.

Sensor principles are currently under investigation for industrial applications. Fritz et al., for example, have investigated the translation of biomolecular recognition into nanomechanical signals. The research was conducted within a collaboration project between IBM Research and University of

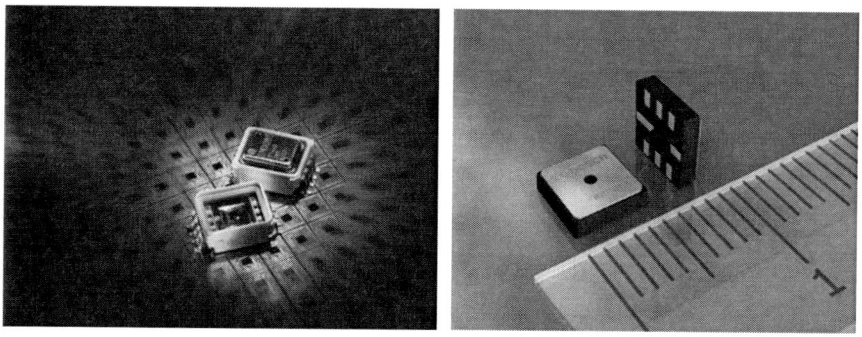

FIGURE 5.2

Packaged micromechanical pressure sensors manufactured by Bosch. Left: SMD085, and right: BMP085 (Photos: Bosch, available online. [7].)

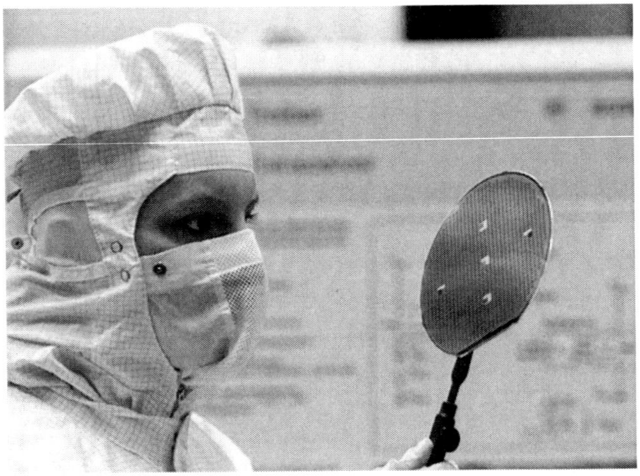

FIGURE 5.3

Wafer production at Bosch in their Reutlingen facility (Germany) under strict cleanroom conditions. (Photo: Bosch, available online. [7].)

Basel, Switzerland, the results of which were published in the prestigious journal, *Science* [8]. They report the specific transduction, via surface stress changes, of DNA hybridization and receptor-ligand binding into a direct nanomechanical response of microfabricated cantilevers. These cantilevers were assembled in the form of an array, and were functionalized with a selection of biomolecules. The differential deflection of the cantilevers was measured as a consequence of molecular recognition during hybridization of complementary oligonucleotides. Other biomolecular interaction studies (protein A-immunoglobulin) have demonstrated the potential for the use of such nanomechanical transducers for the detection of biomolecular recognition. The surface modification method introduced by Shah and Abbott, who investigated the principle for measurement of chemical exposure based on recognition-driven anchoring transitions in liquid crystals, may allow even more smart ways to modify the novel sensor components than those introduced by Fritz et al. [9]. In 2007, Waggoner and Craighead provided a critical review of micro- and nanomechanical transducers for environmental, chemical and biological detection [10].

This work shows clearly that micromechanical pressure sensors made from silicon, by methods adopted from the integrated circuit industry, were just the beginning of this class of devices in the sensor market. We will now discuss application examples which show a potential for optically integrated functions for sensing applications. Micromechanical actuation as a source of excitation in resonant sensors and the principles of actuation in motors, relays, switches, valves and pumps will be covered, from the perspective of specific sensors and actuators in chemical engineering and biology.

5.2 OVERVIEW OF MATERIALS

5.2.1 Single Crystals

Single crystal silicon and silicon-based materials form the largest group of materials used for sensor designers since electronic-compatible production processes are available. Micromechanical sensor designs, however, tend to be much larger in the "footprint" silicon that they use than electronics components, with the smaller number of devices per wafer making production economically questionable. Other materials may be preferred if electronics are not to be integrated into the same platform, or when sensors are intended to operate in harsh environments or at elevated temperatures.

Single crystal quartz is an available, alternative substrate since it has been used routinely for timebase devices in electronics and watches. Gallium arsenide (GaAs) or sapphire (A_2O_3) substrates are used for special applications in the electro-optical domain. They are processed by wet chemical etching which is also used to manufacture diaphragms (often referred to as membranes), beams (called cantilevers in this field), and resonant masses (see Chapter 2 for an illustration of these basic structures in silicon and quartz). These speciality substrates are more difficult to work with than silicon, and to broaden the range of potential designs, dry etch processes have been developed.

These processes are readily available through a diversity of MEMS foundry services or cleanrooms, which are accessible through research centers and universities. So, one could say that a sensor designer is mainly restricted by cost versus performance issues rather than restricted access to an appropriate tool park.

Selection of the best materials and methods for a sensor design may be inherently linked to the background of the designer or the company commercializing the sensing device. The proper operating conditions for the device is the second factor in the choice of materials and methods for fabricating a specific transducer. For chemical sensors, for example, the advantages of integration into silicon are far less obvious, since the devices are often immersed in electrolytes that may short-circuit or shunt the electronics. Sensor packaging is therefore very important, and a substrate that is easier to package than silicon may be preferable. These conditions have driven the need for hybrid integration concepts in the biological and chemical domains, rather than in the mechanical or optical fields. We will return to some of these issues in the discussion of (bio)chemical sensors in Chapter 6.

5.2.2 Amorphous Materials

Micromechanical transducers often incorporate amorphous materials in the packaging of the device. Fused silica and glass are often used in silicon processing lines, and may offer certain benefits, particularly if low thermal conduction and dielectric constants are required in the design. Polymers are also used in this way, and can also actually comprise the sensing or actuator component. Use of these materials potentially reduces the production costs to the order of cents per piece if the device can be made in one injection molding cycle. Limited accuracy, poor machining and the large investments needed to install dedicated tooling equipment have limited the success of such amorphous materials in this area, although some recent investigations into SU-8 photoresist as a mechanical material may change this [11]. Ingrosso et al. have published the results of a functionalization procedure integrated with the fabrication of micromechanical SU-8 cantilevers in order to chemically bind organic-capped iron-oxide (Fe_2O_3) colloidal nanocrystals (NCs) at the photoresist surface, under visible light, ambient atmosphere and room temperature. Their concept resulted in a highly interconnected NC multilayer network, which was demonstrated to be appropriate for real-time detection of acetone vapor [12].

5.3 THICK AND THIN FILM HYBRID MATERIALS

It is difficult to give a defined overview of materials that may be processed by methods as diverse as ceramic tape casting, screen printing, electroplating, lamination, laser machining and all the other techniques in precision engineering. Rapid prototyping print processes or fiber-(electro)spinning, for example, may form excellent materials for advanced micro-nanofunctional systems, but have not been considered due to lack of an identified need. The thin-film community has developed novel fabrication processes for materials that would not exist naturally, because of their hybrid nature or their assembly from fine-tuned layers (so-called multilayer systems). Any of these could be selected as materials in the design of a micromechanical transducer.

On the other hand, the automotive and compliance sensor technology market is highly cost driven, and requires the highest possible sensor performance at lowest manufacturing costs. Here, we can only conclude that with the rising manufacturing standards in the non-silicon industry and upcoming nanotechnology developments for creating novel materials, it is clearly not the integrated silicon industry alone that drives this market sector. We will address the selection of materials and fabrication techniques in the next section, in the context of the working principles of microactuation.

5.4 MICROACTUATION

Table 5.1 presents an overview of structural materials, together with their use and principles of processing and the characteristic properties often explored for microactuation. Any material being considered for implementation into a transducer has its own design rules, as discussed in the literature.

Table 5.1 Overview of materials and application areas

Material	Usage	Process	Characteristic
Ni, Cu, Au	Structure	(Electro) plating	Thick structures
Tungsten	Structure	Thin film	Not attacked by HF
Polyimide	Structure	Thin film	Soft, flexible optical guide
Quartz	Actuation	Anisotropic etching	Piezoelectricity, insulator, transparent
ZnO	Actuation	Thin film	Piezoelectricity
PZT	Actuation	Thick film	Large piezoelectricity
Ti/Ni	Actuation	Thin film	Shape memory effect
TbFe, SnFe	Actuation	Thin film	Magnetostriction
CoNiMn	Actuation	Thin film	Permanent magnet
GaAs	Optics	Thin film	Laser, LED, detector
DLC	Lubrication	Thin film	Low friction and wear
Fluorocarbon	Lubrication	Thin film	Reduce friction and wear
SAM	Lubrication	Thin film	Reduce stiction

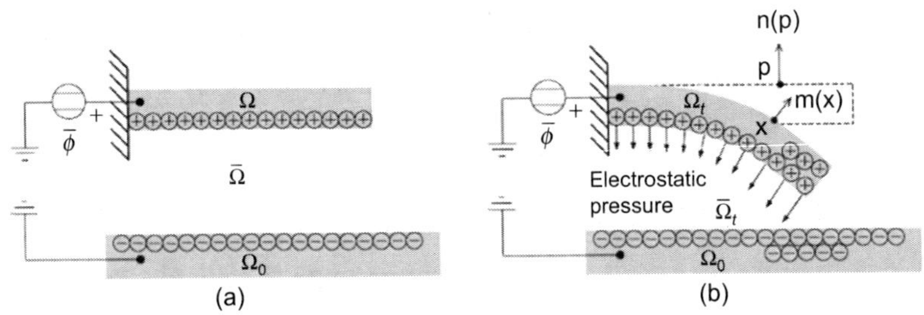

FIGURE 5.4

Illustration of electromechanical coupling in electrostatic MEMS. (A) Device prior to applying electrostatic pressure and (B) devices during applying electrostatic pressure. (Reproduced from [6] with permission from IOP Publishing Ltd.)

One of the working principles often used at the microscale is electrostatic actuation, which, because of the small distance between the capacitor plates (realized by microfabrication approaches), is quite efficient. Oppositely to the efficient function of these small plate distance, it can lead to stiction effects during manufacturing processes and has to be specifically addressed in a process document by a suitable technique, e.g., freeze-drying [13]. Figure 5.4 depicts an example of this electrostatic working mechanism in the actuation of a cantilever, as discussed in Batra et al. [6].

Some materials directly convert the initial source of energy, for example thermal (shape memory alloys) or electrical (piezoelectric) energy, into mechanical movement. Piezoelectricity should be highlighted in MEMS technology. It occurs naturally in certain crystal structures, such as single crystal quartz, but the effect can be created more efficiently in artificial materials, e.g., which exploited the perovskite structure of lead zirconate titanate (PZT). This special material has been used many times in industrial applications. The environmental concerns regarding the use of lead-containing components require alternative piezoelectric materials to be sought. Another reason for the search for lead-free piezoelectric materials are their application at high temperatures.

One other typical microactuation mechanism uses temperature changes to drive mechanical displacement. This can be done with a bimorph (e.g., bimetallic) strip, which uses the mechanical forces caused by the difference in thermal expansion coefficient of the two materials in a thin strip. This difference causes the strip to bend, and can be incorporated into a cantilever design.

Pneumatic expansion and contraction, shape memory or material phase-change caused by the heating or cooling of the system may also be used to create a thermo-reactive transducer, other means of actuation may be magnetic force transduction. An introduction to the design, manufacture and nanoscale engineering of these mechanisms appropriate to an academic undergraduate course is presented in *MEMS and Microsystems* by Tai-Ran Hsu [14].

5.5 PACKAGED SENSORS

Piezoresistive, capacitive and in some cases optical read-out miniaturized pressure sensors make use of a micromachined diaphragm. The capacitor is a simple sensor, made of two parallel plate electrodes. One of the electrodes is also a diaphragm. Diaphragms can be fabricated by silicon etching methods, as discussed in Chapter 2. They are very stiff, and allow controlled displacement of the diaphragm electrode against its counter electrode when a pressure is applied. They are typically a square of 1×1 mm^2 with a thickness of 20 μm. The opposite static electrode is often made directly from the encapsulation material, which is made conductive by a thin film.

The parallel plate structure finds application for pressure sensors, accelerometers and interferometers. Because the capacitances of such sensors are small, stray capacities should be avoided, and the electronic circuitry should preferably be located close to the sensing element. Glass-to-silicon anodic bonding is often utilized as a packaging technique for encapsulation of silicon-electronics. Using the substrate itself as packaging allows encapsulation at the wafer level, thus reducing the costs of batch fabrication and small chip size [15–17]. For the examples depicted in Figure 5.2 in Section 5.1, manufacturers have to consider the assembly at chip (often called *die*) level, device level (packaging into a housing) and the packaging of microsensors at system level. The latter will have to withstand performance testing based on the end-user requirements. A variety of technical solutions were developed for die-level assembly that are considered as CMOS compatible at the back-end of the processing of a sensor. Some processes may still be conducted at the level of the wafer (batch processing), or allow at least easy mounting of the packaged die onto a printed circuit board (PCB). The PCB contains the drive and read-out electronics of the system.

Eventually the individual sensors at the die-level have to be encapsulated by a packaging technique, to protect the delicate mechanical microstructure from the environment. The following subsections summarize key aspects of the packaging process.

5.5.1 From Die to Device Level

A piezoresistive, micromachined, silicon pressure sensor die is often manufactured by a step-and-repeat lithography across the wafer surface. The silicon is etched to create the diaphragm at the wafer level, and the result is bonded to a glass wafer. The individual dies must then be separated into single dies (called dicing), utilizing a circular, diamond, sawing plate. Wafer dicing is carried out on a specially prepared dicing frame (called a chuck) that allows high precision movement of the table from dicing line to dicing line and controlled depth of dicing.

The individual dies are then mounted by pick and place procedures onto the support plate (called a header) that contains the electrical feed-through interconnects and a mounting rim for the casing. Both header and casing are made either from metal or plastic. In some applications, the casing is required to be highly corrosion-resistant or stable at very high temperatures, which demands different packaging solutions.

Plastic encapsulation is commonly used in integrated electronic circuit manufacturing, but no standard packaging procedure applies to all microdevices, and each supply company may have their own favored packaging line. The customer then may reassemble the packaged sensor into a new package, although it may have been better to construct the initial sensor package according to the customer's needs. From the point of view of the mass manufacturing of sensors for different markets, however, the latter is economically more attractive for the supplier.

Figure 5.5 illustrates a cross-sectional view of a die-level package with the electrical interconnects leading to a PCB. Mounting the pins onto an external electronics PCB board, and housing this combination of the electronic board and the packaged die produces a device-level package. At this stage the pressure sensor can be operated as a standalone device, but it is often part of a system controlling an industrial process etc., and therefore needs to be further assembled into a device-level package.

5.5.2 From Device Level to System

One example for a fully packaged system is the *VPFlowScope* made by VP Instruments BV, Delft, The Netherlands. The VPFlowScope combines mass

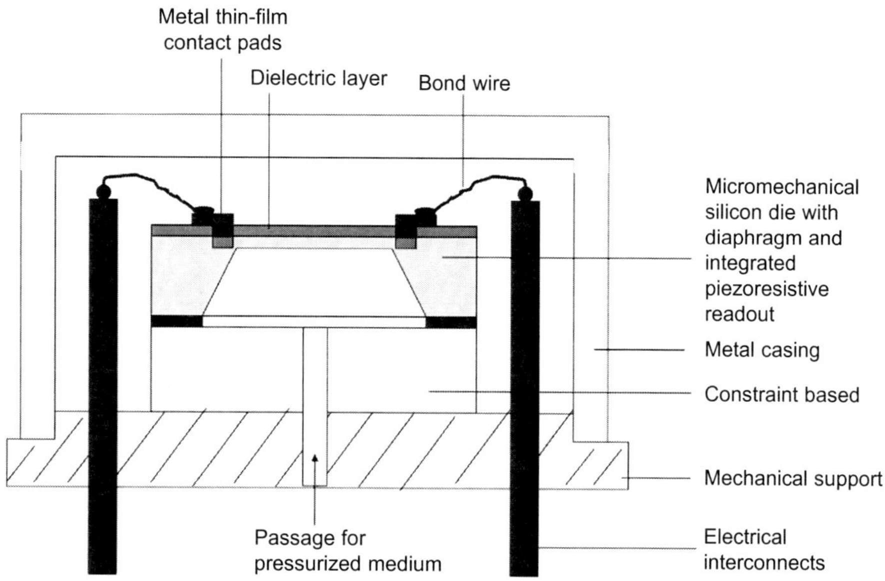

FIGURE 5.5

Assembly of a die into a package. (Figure is drawn based on examples presented by Hsu [14].)

flow, pressure and temperature sensors into one measurement system, combined with a built-in data logger. The system-level package is compatible with a special installation procedure, called hot tap drilling, which allows installation points to be created in compressed air piping without interrupting the compressed air supply. This system level package makes the VPFlow-Scope easy to install and to use as an all-in-one instrument for compressed air measurement and similar applications.

Compressed air systems are installed to support many different types of industrial production processes. The users of such systems often rely on the proper function of the compressed air pipes, however, leakage, misuse or other reasons for non-optimal configurations in the piping system may lead to additional costs. Using a flow-management system may give cost savings of up to 50% of the current operation. Although the pressure sensor integrated with the VPFlowScope is not based on a capacitive diaphragm, as discussed above, it utilizes silicon micromachining in its pressure and flow sensing elements with smartly integrated electronic temperature compensation and

FIGURE 5.6

Example of an assembly of different sensor devices into one user-proof system package ready to use by the customer: VPFlowScope installation starting kit (left), a typical operating environment (middle) and the installed VPFlowScope (right). (Image courtesy: VPInstruments BV. Figures available online. [18].)

is therefore a good example of the micro system-in-a-package approach. Figure 5.6 shows the system components as a starting installation kit (left), a typical installation environment of a compressed air system (middle) and a close-up of a VPFlowScope installed into a pipe line (right). More information may be obtained from the company's website [18]. The measurement principles and design considerations of the type of devices adopted in the product range of VPInstruments have been presented by van Putten et al. [19].

5.6 SILICON AS A MECHANICAL MATERIAL IN RESONANT MICRODEVICES

Silicon and the associated integrated circuit fabrication techniques have been the most investigated material and fabrication methodologies for mechanical sensors for the last three decades. This can be largely explained by the success of silicon in the batch fabrication of electronic devices. There is as yet no reasonable replacement as a handling wafer if one considers that some electronics may be integrated into an innovative sensor. Industrial interest mainly centers on redundancy and organizing multiple sensors in one platform. The benefits of using silicon have clearly sustained research and development by various successful transitions to industry for applications employing temperature sensors, pressure sensors and accelerometers. One special class of sensors works at resonance, whereas the principle of measurement relies on a frequency-shift, providing an output signal very attractive for precision measurements as well as new implementation areas, such as for example measuring molecular interactions [8, 20]. The introductory section of the publication by Tilmans et al. summarizes some key aspects and developments concerning resonant devices, so the next subsection uses their wording [21]. Tilmans, who works at the respected IMEC Institute, Leuven, Belgium, is an expert in this field, and has successfully translated these types of devices to industry for many years. Other overviews of this field of research are the paper on "Resonant silicon sensors" by Stemme published in 1991 [22], the comprehensive *Handbook of Sensors and Actuators*, in Volume 8 of which (edited by Middelhoek), Bao discusses micro mechanical transducers [4, 23]. Elwenspoek and Wiegerink have presented a book specifically on *Mechanical Microsensors*, which covers the essential development stages, design and applications of resonant microdevices [3].

5.6.1 Resonant Sensors

Resonant sensors provide, in addition to their excellent stability, resolution and accuracy, a signal output in frequency form that is immune to intensity fluctuations, and allows the devices to be easily connected with digital systems, which is required for its effective employment as a measurement device. The structure should only respond to changes in the load to be measured. We have already briefly discussed the pressure sensor in Section 5.5.1. One of the first resonant strain gauge-type pressure transducers was described by Belyaev et al. in 1965 [24]. In their design, the strain gauge is mounted on top of a diaphragm by means of two brackets. Similar designs were also introduced by Greenwood [25] and Thornton et al. [26]. Silicon bulk micromachining was used to fabricate these devices. Ikeda et al. presented a pressure sensor design in which the resonators are held in evacuated microcavities on top of the diaphragm. Operating in a vacuum enhanced performance tremendously, gaining a so-called higher Q-factor (Q = quality) [27]. This group used the single-crystal silicon, considered to be a classical construction material, in combination with selective epitaxial growth techniques and a high boron etch stop for the fabrication of the device by etching in ethylendiamin perizine (EDP). The latter is no longer used in production environments. An alternative way of fabricating a sealed resonator device was reported by Guckel et al. [28], who used the deposition of a fine-grained polysilicon layer by LPCVD and sacrificial layer etching to construct their device. Both technologies are very attractive because of their ability to manufacture in batches. We will discuss some of these aspects in more detail in the following section, but focus here on silicon as a mechanical material. Because most device designs for sensor systems are inherently difficult to comprehend, we will first give a design example for the application of silicon as a mechanical material, by describing the case of a three-sides clamped diaphragm.

5.6.2 Diaphragms as Micromechanical Couplers

Although the structure discussed here was originally designed as an actuator for a micro-optical scanning device, it may also be utilized in the design of a resonant cantilever-type sensor. Diaphragms are clearly one of the basic building blocks in micromechanical sensor and actuator design, and their different resonant behavior with respect to clamping is an interesting field of research.

Figure 5.7 depicts a diaphragm with an integrated boss-feature (also called a mesa). This mesa allows the micromechanical cantilever to be connected to the microactuator at a later processing stage. The diaphragm has been etched in a KOH bath at the wafer-level, and at the back of the wafer a thin film of gold has been deposited to create a conductive layer. Subsequently the wafer has been diced and the diaphragms mounted on a frame, which is shown in Figure 5.8. To excite the micromachined structure, a commercial piezoelectric PZT-disc has been glued to the back of the diaphragm using silver-epoxy,

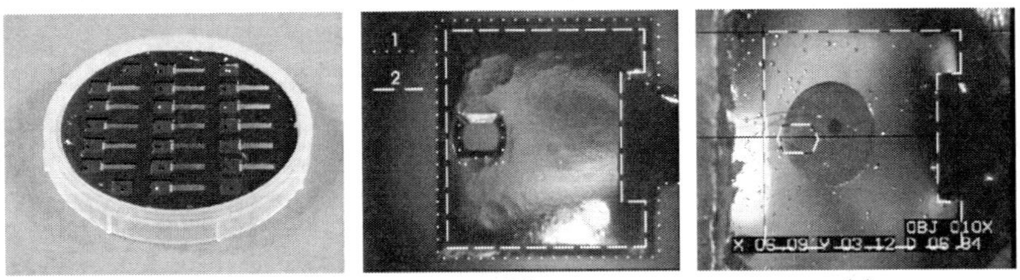

(a) (b) (c)

FIGURE 5.7

Silicon diaphragm fabricated by silicon micromachining. A commercial piezo-element is mounted to the diaphragm applying a mechanical excitation source by a small sinusoidal drive voltage to the piezo, which will subsequently drive the membrane into resonance at its eigenvalues. View at wafer-level (a), diaphragm top-view with mesa at the left-hand side of the diaphragm (b) and mounted piezo-disc at the back of the diaphragm, dashed lines mark the diaphragm area (c). (Reproduced from R. Luttge, PhD Thesis, 2003, University of London, UK.)

(a) (b)

FIGURE 5.8

Mounted diaphragm on aluminum support frame including a view with a hybrid-integrated glass cantilever component bonded to the mesa-structure of the diaphragm, whereas the components act together as barcode reading scan head device, (a) and a view onto the connected piezo-disc at the back of the diaphragm with PCB and cable connector (b). (Reproduced from R. Luttge, PhD Thesis, 2003, University of London, UK.)

which connects the piezo-disc electrically to the gold layer and to the frame, respectively. The opposite side of the piezo-disc is wire-bonded to an isolated printed circuit board (PCB) that was also mounted to the frame using epoxy glue. Standard electrical cables can then be used for the connections from the frame and the PCB to the frequency generator providing a sinusoidal output voltage to the piezo-disc for excitation.

Figure 5.8 shows the completed microelectro(opto)mechancial system, which includes a glass cantilever for the integration of an optical function into the system (not further discussed here), all mounted on a conventionally milled aluminum frame for testing the device on an optical bench. In this specific design, the wafer-level manufactured diaphragm is intended to provide micromechanical coupling between a commercially available non-CMOS compatible PZT element and another non-CMOS compatible optical integrated cantilever device. The micromachined silicon diaphragm can also function as a platform for full electronic integration of a micro-opto-electromechanical system (MOEMS) (for more details on MOEMS, see Section 5.7.1 below, and also Chapter 3, Section 3.5.2).

Figure 5.9 shows the first five eigenvalues of a three-sides clamped diaphragm, assuming an idealized design without the coupled mass of the mesa, the piezo-disc or the cantilever. Knowledge of the resonant frequency

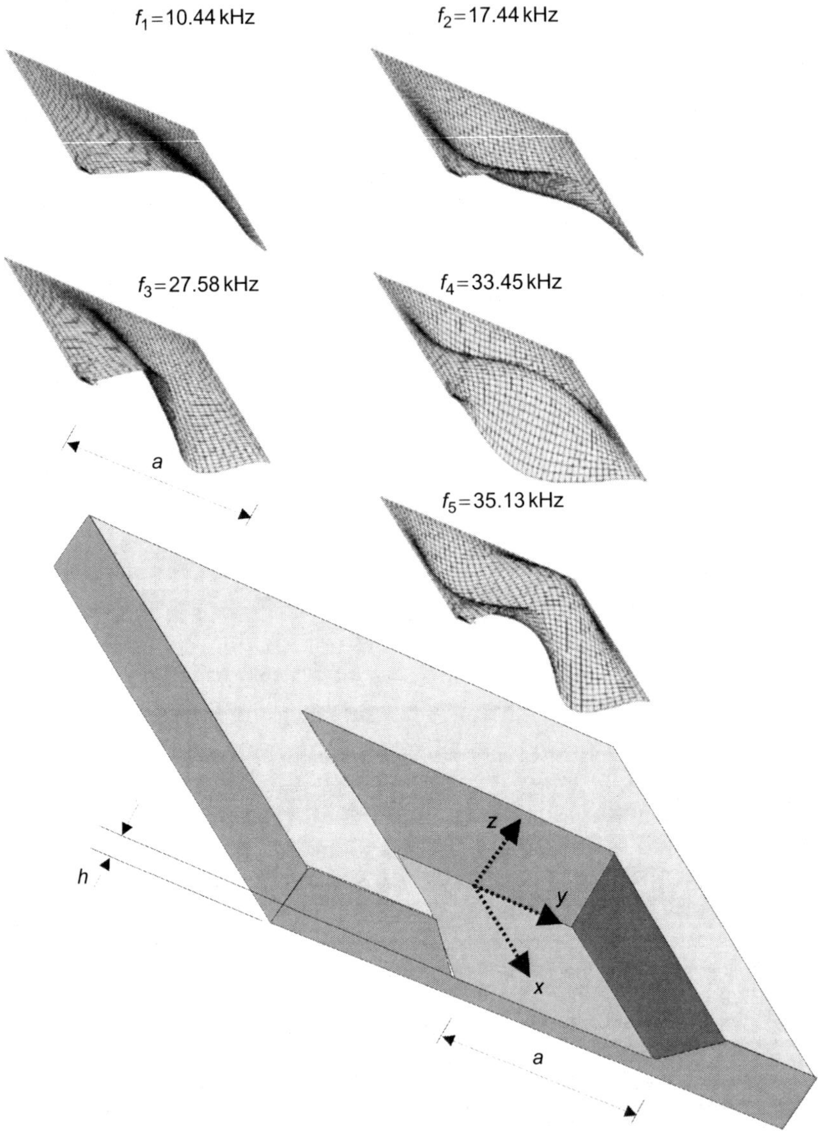

FIGURE 5.9

Visualizing the eigenvalues or mode shapes of a diaphragm clamped at three sides. Mode shape surface plots generated numerically by E. Oosterbroek for a rectangular silicon diaphragm with side length $a = 6.5$ mm and a diaphragm thickness $h = 45\,\mu$m. (Reproduced from R. Luttge, PhD Thesis, 2003, University of London, UK.)

is a key parameter in the design optimization of a resonant system. Of course the damping of the system also plays an important role. Excessive damping would prohibit resonance, and too little might damage the system. For micromechanical cantilevers, large quality factors have been achieved (1000 or higher), therefore the damping ζ can have very small values while critical damping occurs at $\zeta = 0.2$.

5.7 INFORMATION SOCIETY

Microtechnology not only offers a reduction in device size, but also enables cost-efficient fabrication via the parallel output of many components – so-called batch processing. Once it has been determined that a miniaturized system would enhance performance, and be economically viable in large-volume production, this parallel processing approach will then repay the initial (large) investment. The overview of studies in this chapter are an introduction to the paradigm shift of industrial applications caused by the evolution of microfabrication technology. All engineering disciplines have benefitted from previous developments in integrated circuits processing techniques, and this has fed into applications by inter- and multidisciplinary research, which have yielded novel fields of systems design such as MOEMS, BioMEMS and microreactors.

All these different fields have specific materials, requirements and robustness, sensor selectivity and sensitivity, as well as accuracy and precision specifications. Sensors in the field of BioMEMS and microreaction technology will be discussed in more detail in Chapters 6 and 7, and the following subsections cover a variety of cases which are either fully commercialized or investigated thoroughly at component level from a market perspective. Information technology (IT) is still the most advanced commercial user of sensor applications. However, all other sectors are benefitting from the early-adopter position of IT, and follow the trend of the *More-than-Moore* concept, which further promotes the systems-on-a-chip approach for micro-nanomechanical sensor systems.

5.7.1 Micro-Opto-Electromechanical Systems

Optical engineering consists of the fabrication and assembly of optical components, such as lenses, mirrors, prisms, and refraction and diffraction elements. One could include the mechanical features of waveguiding applications such as light distribution, signal modulation and retrieval in this category. It is possible to emulate some of these optical components at the micro domain, but the features that move along a two-fold axis (micromirror) and the passive alignment of components remain as critical design problems. Fiber couplers and out-of-plane mirror elements are examples of devices which face these limitations. Although at the end of the 1990s there was a major economic dip in the telecommunication market because of the *dot com meltdown*, still MOEMS industrial developments and cutting-edge system design research has progressed. Some of the first successful industrial products spinning off the continuous developments in MOEMS were portable point-of-sales barcode readers (Figure 5.10), text-recognition and advanced e-reading headset systems (Figure 5.11), and micro-mirror integrated digital projectors and displays (Figure 5.12).

Barcode readers are used for in-store inventory, document management, vehicle registration, etc. So they need to be lightweight and portable. Examples of such systems are presented in Figure 5.10. The technology for barcode reading has applications in a variety of industries, such as healthcare, mobile data collection and manufacturing. Portable systems may be based on different technologies including contact scanning, CCD imaging and laser scanning, of which the systems depicted in Figure 5.11 give some examples. In general, laser scanning is achieved by beam deflection using

FIGURE 5.10

Datalogic Magellan 8500XT retail barcode scanner (left, source: Datalogic, reproduced from image accessed via online distributor site, 2011), hand-held scanner (middle, source: Symbol Technologies, reproduced from image accessed via online distributor site, 2003) and integrated wearable WT4000 (Motorola) scanner series on finger, which allows use of both hands for other actions (right, source: Symbol Technologies, reproduced from image accessed via online distributor site, 2003.)

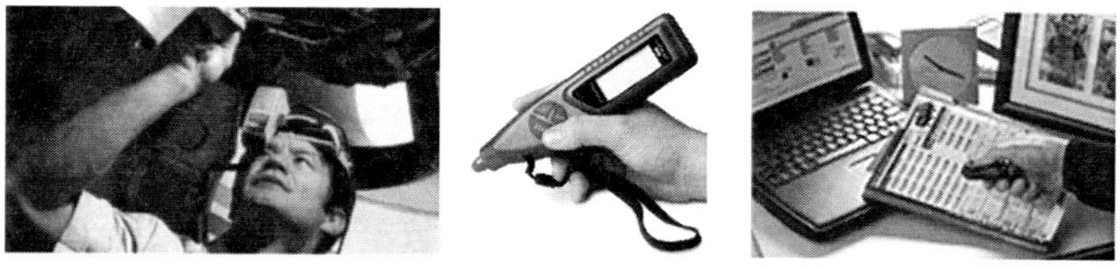

FIGURE 5.11

Diversity of scanning tools. Head-mounted retina display (left), hand-held text scanner (middle) and scanner for recognition of barcodes (right). (Reproduced from images accessed via online distributor sites 2003.)

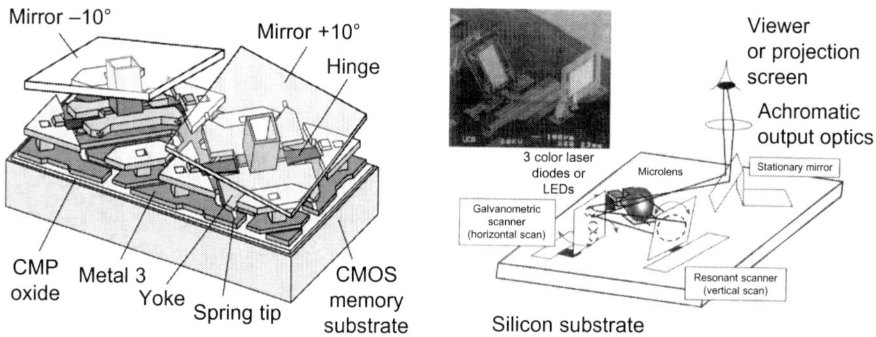

FIGURE 5.12

MEMS-based digital mirror projection display (left, reproduced from [32] with permission ©IEEE 1998) and another type of display device configuration (right, reproduced from reference [33].)

a galvo-mirror, with the returned light being collected and focused onto a photodiode by a parabolic mirror.

New systems in micro-optics rely on surface micromachining, deposition and patterning of polysilicon in a sequence with sacrificial layers to obtain movable structures. We discussed the principle of surface micromachining in Chapter 2, Section 2.3.2. Examples of these micromachined optical-integrated systems are illustrated in Figure 5.12. One approach in micro-optics is to redirect light by actuated mechanical carriers known from switching devices [29, 30], but this does not incorporate an integrated waveguide. Also, the advanced waveguide switching devices that do exist lack adequate displacement range. An alternative approach miniaturizes the conventional optical bench, which leads to a large number of very impressive scanning display devices using resonant mirror components and comb drives to actuate them. Tien et al. have published examples of such micro-optical benches for scanning devices. These are depicted on the right-hand side of Figure 5.12. This type of device requires an increased number of total components, which may be not acceptable for certain production lines. Over the years, surface micromachining technology has matured commercially. The service is offered by MEMSCAP, Inc., a successful French MEMS service supplier. The cost for a multi-user MEMS process (MUMPs) using either the polysilicon, metal or SOI process line with a standard die site on the wafer of 1 cm × 1 cm and a total of 15 unreleased die delivered to the client is 3700 US-dollars/run for academic use and 5300 US-dollars for non-academic use. [31]

Power splitters, and more advanced systems such as waveguides, optical amplifiers and modulators, have been fabricated for telecommunication devices and photonics. Wu et al. present an overview of optical MEMS technology [34]. A marriage between previously developed processes for microelectromechanical systems and optical systems has led to the emergence of the research field of micro-opto-electromechanical systems (MOEMS).

When I conducted my PhD research at the Optical and Semiconductor Devices Group, Imperial College, London, UK, I was part of the research team of the *Optical Scanner Project* headed by Prof. Syms from the years 1999 to 2003. In this context, I wish to acknowledge funding by the EPSRC (British Research Council). In the realm of the many market opportunities associated with integrated optics in the late 1990s, I investigated fabrication technologies for optical scanners based on micromachined cantilevers. This *Resonant Cantilever Optical Scanner* (RCOS) is an example of the application of MOEMS technology.

One of the objectives in the design of a barcode reading scan head device (depicted in Section 5.6.2, Figure 5.8) is to allow the integration of electronic components in the system prior to the back-end MST processing of the diaphragms. For aluminum integration, for example, the thermal budget must be lower than 450°C. The MST processing steps have therefore to be chosen appropriately to avoid destroying the electronics previously realized in the chip. Besides this obvious need for strict temperature control, micromachining in SC-quartz, for example, should also stay below 500°C because otherwise a transition from piezoelectric α-quartz to non-piezo β-quartz occurs, which would jeopardize the prospective of a high integration density.

Multiple functions within same material (monolithic integration) would be no longer attainable. This section will therefore quote some examples from the literature which are relevant to the investigation of process technology for optical films at a modest temperature range.

We have already discussed various optical film deposition techniques and their impact on PLCs in Chapter 3. The examples of sol-gel based PLC, epitaxial deposition and sputtered optical layers complete this overview. High quality and relatively low manufacturing costs of spin-on glasses makes them attractive for the mass fabrication of devices. Passive sol-gel based integrated devices have already been demonstrated in earlier research, for example, at the Optical and Semiconductor Devices Group [35–38].

Cheng et al. [39] described a spin-on-glass process that simplifies the usually much more complex PVD or CVD deposition of lithium tantalate ($LiTaO_3$) layers. This enables many applications that use piezoelectric, nonlinear optical, electro-optical and acousto-optical properties within MOEMS. Lithium tantalate is spin-coated under a nitrogen atmosphere from a solution of $LiOC_2H_5$ in ethanol with $Ta(OC_2H_5)$ onto silica-on-silicon wafers. The paper gives a detailed characterization of multilayer thick films resulting in a root-mean-square surface roughness of $Ra = 6.3$ nm and a refractive index of $n = 1.84$ measured at the 633 nm wavelength. However, elevated temperatures of 600 °C were needed [39].

Plasma enhanced CVD is an attractive alternative technique for deposition at a lower temperature range, which can be used to create MOEMS fiber-to-waveguide couplers. Typically, deposition takes place between 200°C and 350°C. Figure 5.13 shows a device that is based on silicon-oxy-nitride (SiON)-on-silicon technology [40]. To obtain a high coupling efficiency it is important that the optical mode profile of fiber and planar guide are matched [41]. This can be achieved either by a lens, prism, grating or a tapered end of the waveguide. When using a tapered end, as suggested by de Ridder et al., fabrication techniques must be used that have tight control of tolerances, to achieve a very sharp point (taper radius smaller than 500 nm) for optimal optical performance [40]. The proposed design, shown in Figure 5.13, can be realized with only moderate demands on the quality of the photolithographic pattern definition. A so-called spot-size transformer for fiber-to-chip coupling is presented. This example of SiON deposition with a refractive index of $n = 1.7$ and a thickness of 300 nm for single mode propagation can be also utilized in RCOS design, which has been briefly mentioned above.

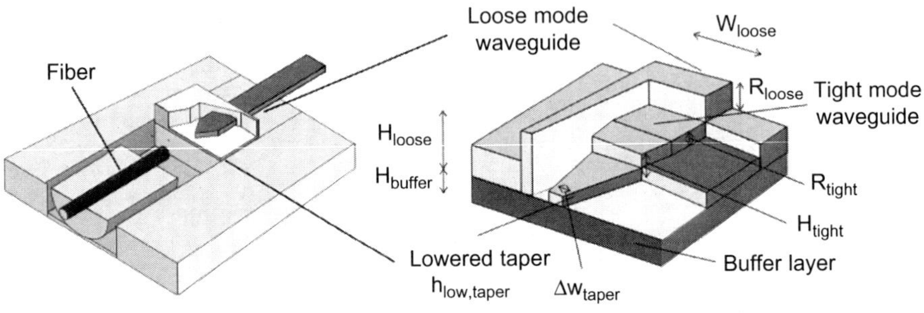

FIGURE 5.13

Optical integrated device for fiber-chip coupling using SiON/SiO_2 waveguide technology. (Figure redrawn from reference [40].)

Another possible coating technique for a silicon MEMS-based integrated optical device is thin film magnetron sputtering of Al_2O_3. During magnetron sputtering, slab waveguides can be deposited from an Al_2O_3 target in an argon gas atmosphere containing some oxygen. A photoresist mask can be applied to the layer and subsequently the Al_2O_3 slabs can be patterned by accelerated argon ion milling [42]. This approach results in low roughness along bends. For telecommunication applications the quality of the material must be enhanced by an annealing step at approximately 700°C or higher. However, the magnetron sputter technique could be considered without the additional temperature treatment for less demanding applications, similar to the above mentioned RCOS for barcode reading. M.K. Smit developed this process during his PhD project at the Delft University of Technology, The Netherlands. His thesis also details many other materials and fabrication aspects involved in the design of integrated optical applications. He describes functional optical designs and gives a comprehensive overview of the theoretical background of waveguides, including computation for bends, couplers and optical phased arrays. Essential parameters and measurement methods for the characterization of waveguides are also given [42]. A variety of textbooks and review papers provide a good starting point for advanced reading on MOEMS and integrated planar optics [34, 43–45].

PECVD is a potential alternative to the magnetron sputter technique for making low-temperature optical slab waveguides. Kim et al. [46] have investigated the microstructure of this type of film. Figure 5.14 depicts a cross-section of an AlO_x film on silicon. Low-loss alumina waveguide

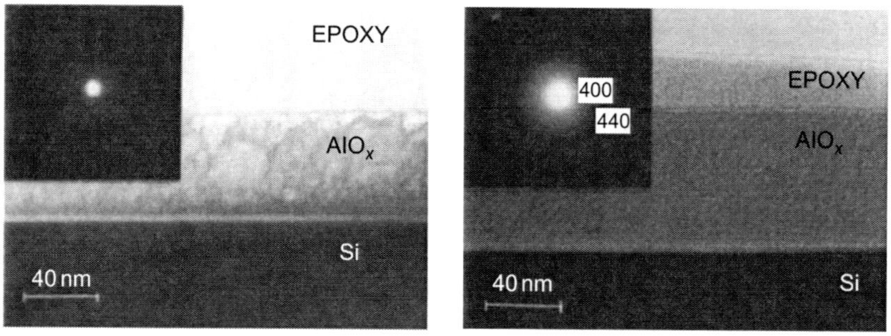

FIGURE 5.14

Cross-sectional TEM micrographs and diffraction patterns of aluminum oxide films deposited at 120°C (left) and 250°C (right). (Reproduced from reference [46].)

fabrication by atomic layer deposition (ALD) suitable for the ultra-violet and visible spectral regions has been recently reported by Aslan et al. [47]. ALD films were deposited at 200°C and 300°C, and are potentially useful for several integrated optical devices, which may include post-CMOS integrated applications.

Until the early 1990s, optical integrated circuits (OICs) concentrated on monolithic techniques to confine light into a guiding region for mixing or splitting, with integrated laser technology being used to make compact laser diodes, or to develop other opto-electronic devices in a variety of semiconductor materials. Typical techniques for the fabrication of waveguides are derived from thin-film processes, such as physical and chemical vapor deposition and subsequent patterning by photolithographic means. The use of thick oxide layers (2–20 μm) as an optical buffer layer on silicon wafers is one of the most widely used techniques for telecommunication applications, where the wavelength is in the micrometer range. The optical buffer layers are applied to the silicon by flame hydrolysis, which results in highly stressed silica. This is not suitable to be released from the silicon substrate, hence both the silica and the substrate are lost to further micromechanical usage in MOEMS. Nevertheless, thick buffers are needed to minimize the absorbance of light into the silicon substrate. The techniques addressed above may already present a solution to monolithic integration of the RCOS. However, when silicon micromachined cantilevers are employed in the RCOS design, thick claddings would bc unavoidablc, which suggcst that thc wavcguidc-cladding-cantilever microstructure could not be released from the substrate.

Since a free-moving cantilever with integrated optical function is the aim of the project, other means of integration technology must be found.

To overcome such problems is standard industrial practice, hybrid solutions for optical device integration are frequently sought. If the two process lines for electronic and micro-opto-mechanical techniques are strictly separated, production of an integrated RCOS becomes much more feasible by using already established foundry services for the realization of the required functions. Only the final packaging steps of combining the opto-electromechanical integrated devices by means of subsequent processing of the MEMS/MOEMS structure onto the electronic component wafer would require a low thermal budget.

With respect to RCOS, Roberts et al. have presented some results based on precision engineered scan units [48, 49]. A light-guiding optical fiber served as a cantilever and was mounted on a commercial piezoelement actuator. Work performed by Roberts et al. led to the conclusion that fine-mechanical assembly of components, even on a cm-scale, is not sufficiently reproducible for a production process. A reduction of device size, and eventually spot size, is expected to enhance scanner performance, which allows for new applications to be developed. The demands for optical scanning in the fields of medical endoscopy, display accessories embedded in lightweight head mounts or even cosmetic spectacles, are a driving force for the continuous reduction of device dimensions. The realization of precisely directed illumination from a single mode integrated waveguide in a variety of wavelengths could lead to many other devices, for example, in metrology, ophthalmology or other medical applications [50–52].

5.8 CONCLUSIONS

The main aim of this chapter is to address the challenge of integrating skills and technologies from different scientific disciplines onto one chip. The demand of pure monolithic integration in one and the same material (mainly silicon) places limitations on the system. The engineer always has to compromise between the material properties and its fabrication capabilities. Development efforts should be focused on new applications, which combine optimized microtechnological feasibility with the best choice of material. For example, a variety of non-linear optical effects can only be produced with certain opto-active materials. To manufacture new products, new technological combinations should be explored, geared towards high efficiency. Although silicon is an excellent MEMS material, researchers have meanwhile also developed batch fabrication processes for other materials (see, for example, the processes described in Chapter 3). Section 5.2 has given an overview of different materials which are used in micromechanical sensors and actuator designs. However, continuous development in defining and shaping novel materials is required in order to meet the demands of a new generation of application fields. These different materials also require different mounting techniques, for example, index matching adhesives and other bond mechanisms were developed to meet technical and optical requirements [53].

Development strategies within the information technology industry are mainly based on functions in the electrical, magnetic or optical domain, and their combinations. Most techniques, from top-down approaches to the manufacture of sub-micron features described in Chapter 2, are generally applicable for integration in all the different energy domains. However, as we discussed in this chapter, the optimum choice of materials and process depends on the function of a specific system or component, the existing production capabilities and the product-to-market combination.

Microelectronic devices already regulate, control, enhance and dominate daily life. Think of your overfull e-mail or voice mail box in this context. Furthermore, the effectiveness of a modern city depends entirely on computerized technology for transport, finance and other customer services (e.g., banks, insurances, health care services). Think about all the plastic cards in your purse that give you access to these digital networks. The application of microsystems technology helps to save resources and to enhance operational performance. As in other fields, optical instrumentation went through a process of miniaturization, too, utilizing micromechanical transducers. Many new application fields are being explored at the investigator level, e.g., in the context of nanomechanical transducers for the identification of biological recognition processes, and are all intended to improve human life and the environment. If one does not yet know the specific answers that address an immediate problem, gathering information (sensing) is the first step towards innovation in any inquiry.

Specifically with respect to technological advances in the context of microfabrication for industry, microsystems technology (MST) has found that certain prohibiting effects on a macroscalc can become useful for engineering at the micro level, e.g., the effect of resonance discussed above. MST combines the fabrication of components to exploit effects on the microscale with all other factors of systems design known from the conventional engineering approach. Furthermore, the process of miniaturization leads to new discoveries, such as the use of photonic crystals to manipulate light [54, 55].

When optical strategies were combined with microelectronics, the optical integrated circuit (OIC) was developed, based on optical waveguides instead of electrical conductors (see also Chapter 3, Section 3.6) [56, 57]. OICs are based on similar fabrication techniques to microprocessors, as they have been used for the manufacture of microprocessors. Their production uses thin-film integration concepts and in-plane operations, hence an optical integrated device is also often called a planar light circuit (PLC). In early efforts, which relied on planar techniques, mostly monolithic (i.e., all components fabricated on/in the same substrate) integrated approaches were pursued. However, the field of research and applications has grown so fast that existing technologies have not always been capable of delivering fully monolithic integrated devices with the desired level of performance. Also, niche markets need to use hybrid devices to speed up the development cycle and lower the cost of technical solutions by using standardized pre-packaged components. Powering a new generation of optical networking will require process and device synthesis on a much higher level of systems integration than is currently possible. The reliable demonstration of the merging of micromechanical and optical systems is a model case for further innovation by integration of micromechanical sensors and actuators with higher level architecture systems.

References

[1] A.F.P. van Putten, *Electronic Measurement Systems*, Prentice Hall, Englewood Cliff, NJ, ISBN 0-13-251-885-6, 1988.

[2] J. Wilson, *Sensor Technology Handbook*, Elsevier, Newnes, Oxford, ISBN 0-7506-7729-5, 2004.

[3] M. Elwenspoek, R. Wiegerink, *Mechanical Microsensors (Microtechnology and MEMS)*, Springer, Berlin, ISBN 3-540-67582-5, 2001.

[4] M.-H. Bao, *Micro Mechanical Transducers: Pressure Sensors, Accelerometers and Gyroscopes. Handbook of Sensors and Actuators*, vol. **8**, (edited by S. Middelhoek) Elsevier, Oxford, ISBN 978-0-444-50558-3, 2000.

[5] W.P. Eaton, J.H. Smith, Micromachined pressure sensors: review and recent developments, *Smart Mater. Struct.* **6** (1997) 530–539.

[6] R.C. Batra, M. Porfiri, D. Spinello, Review of modeling electrostatically actuated microelectromechanical systems, *Smart Mater. Struct.* **16** (6) (2007) R23–R31.
[7] http://www.bosch-press.de (accessed 07.04.11).
[8] J. Fritz, M.K. Baller, H.P. Lang, H. Rothuizen, P. Vettiger, H.-J. Meyer, et al., Translating biomolecular recognition into nanomechanics, *Science* **288** (2000) 316–318.
[9] R.R. Shah, N.L. Abbott, Principles for measurement of chemical exposure based on recognition-driven anchoring transitions in liquid crystals, *Science* **293** (2001) 1296–1299.
[10] P.S. Waggoner, H.G. Craighead, Micro- and nanomechanical sensors for environmental, chemical, and biological detection, *Lab. Chip.* **7** (2007) 1238–1255.
[11] M. Calleja, M. Nordström, M. Álvarez, J. Tamayo, L.M. Lechuga, A. Boisen, Highly sensitive polymer-based cantilever-sensors for DNA detection, *Ultramicroscopy* **105** (2005) 215–222.
[12] C. Ingrosso, E. Sardella, S. Keller, S. Dohn, M. Striccoli, A. Agostiano, et al., Surface functionalization of epoxy-resist-based microcantilevers with iron oxide nanocrystals, *Adv. Mater.* **22** (30) (2010) 3288–3292.
[13] S. Deladi, V. Svetovoy, G.J.M. Krijnen, M.C. Elwenspoek, Flash release – an alternative for releasing complex MEMS devices, *J. Micromech. Microeng.* **14** (12) (2004) 1659–1664.
[14] T.-R. Hsu, *MEMS and Microsystems*, Wiley, Hoboken, NJ, ISBN 978-0-470-08301-7, 2008.
[15] M. Esashi, Encapsulated micro mechanical sensors, *Microsyst. Technol.* **1** (1) (1994) 2–9.
[16] M. Esashi, Wafer level packaging of MEMS, in: *TRANSDUCERS 2009 – 15th International Conference on Solid-State Sensors, Actuators and Microsystems*, Denver, Colorado, USA, 2009, pp. 9–16.
[17] M. Esashi, Micro/nano electro mechanical systems for practical applications, *J. Phys.: Conf. Ser.* **187** (art. no. 012001) (2009) pp. 1–10
[18] VP Instruments. www.vpinstruments.com, (accessed 07.04.11).
[19] A.F.P. Van Putten, M.J.A.M. Van Putten, M.H.P.M. Van Putten, Silicon thermal anemometry: Developments and applications, *Meas. Sci. Technol.* **7** (10) (1996) 1360–1377.
[20] R.M. Langdon, Resonator sensors – a review, *J. Phys. E: Sci. Instrum.* **18** (2) (1985) 103–115.
[21] H.A.C. Tilmans, S. Bouwstra, J.H.J. Fluitman, S.L. Spence, Design considerations for micromechanical sensors using encapsulated built-in resonant strain gauges, *Sens. Actuators: A Phys.* **25** (1–3) (1990) 79–86.
[22] G. Stemme, Resonant silicon sensors, *J. Micromech. Microeng.* **1** (1991) 113–125.
[23] P. Regtien, (Ed.), *Handbook of Sensors and Actuators* (Series), Elsevier, Oxford, Measurement Science for Engineers, ISBN 978-1903996584, 2004.
[24] M.F. Belyaev, D.D. Dorzhiev, L.G. Etkin, Vibration-frequency pressure transducer, *Instrum. Constr.* **10** (1965) 10–13.
[25] J.C. Greenwood, Etched silicon vibrating sensor, *J. Phys. E: Sci. Instrum.* **17** (8) (1984) 650–652.
[26] K.E.B. Thornton, D. Uttamchandani, B. Culshaw, Novel optically excited resonant pressure sensor, *Electron. Lett.* **24** (10) (1988) 573–574.

[27] K. Ikeda, H. Kuwayama, T. Kobayashi, T. Watanabe, T. Nishikawa, T. Yoshida, et al., Silicon pressure sensor integrates resonant strain gauge on diaphragm, *Sens. Actuators: A Phys.* **21** (1–3) (1990) 146–150.

[28] H. Guckel, J.J. Sniegowski, T.R. Christenson, F. Raissi, The application of fine-grained, tensile polysilicon to mechanically resonant transducers, *Sens. Actuators: A Phys.* **21** (1–3) (1990) 346–351.

[29] M. Hoffmann, P. Kopka, E. Voges, Optical fibre switches based on full wafer silicon micromachining, *J. Micromech. Microeng.* **9** (1999) 151–155.

[30] R. Conant, J. Nee, K. Lau, R. Muller, Dynamic deformation of scanning mirrors, *EEE/LEOS Conference on Optical MEMS*, Kauai, Hawaii, August 20–24, 2000, pp. 49–50.

[31] Memscap Mumps pricing. http://www.memscapinc.com/nc-pmumps.pricing.html, 2011 (accessed 07.04.11).

[32] P.F. Van Kessel, L.J. Hornbeck, R.E. Meier, M.R. Douglass, A MEMS-based projection display, *Proc. IEEE* **86** (8) (1998) 1687–1704.

[33] N.C. Tien, O. Solgaard, M.H. Kiang, M. Daneman, K.Y. Lau, R.S. Muller, Surface-micromachined mirrors for laser-beam positioning, *Sens. Actuators A* **52** (1–3) (1996) 76–80.

[34] X.-F. Wu, Y.A. Dzenis, Guided self-assembly of diblock copolymer thin films on chemically patterned substrates, *J. Chem. Phys.* **125** (17) (2006) 174707.

[35] R.R.A. Syms, V.M. Schneider, W. Huang, M.M. Ahmad, High-delta *n* silica-on-silicon channel waveguides based on sol-gel germanophosphosilicate glass, *Elect. Lett.* **33** (1997) 1216–1217.

[36] F. Horowitz, E.J.C. Dawnay, M.A. Fardad, M. Green, E.M. Yeatman, Towards better control of sol-gel film processing for optical device applications, *J. Nonlinear Opt. Phys. Mat.* **6** (1997) 1–18.

[37] V.M. Schneider, R.R.A. Syms, Gallium-doped sol-gel glass waveguides on SI substrates by SC-RTA, *Elect. Lett.* **34** (1998) 1849–1850.

[38] E.M. Yeatman, M.M. Ahmad, O. McCarthy, A. Martucci, M. Guglielmi, Sol-gel fabrication of rare-earth doped photonic components, *J. Sol-Gel Sci. Tech.* **19** (2000) 231–236.

[39] S.-D. Cheng, Y. Zhou, C.H. Kam, Y.L. Lam, Y.C. Chan, W.X. Que, Sol-gel derived thin films of $LiTaO_3$ on SiO_2/Si substrates for optical waveguide applications, *Fiber Integr. Opt.* **20** (2001) 45–52.

[40] R.M. de Ridder, R.A. Wijbrans, H. Albers, J.S. Aukema, P.V. Lambeck, H.J.W.M. Hoekstra, A spot-size transformer for fiber-chip coupling in sensor applications at 633 nm in silicon oxynitride, *LEOS95 Annual Meeting*, San Francisco, 1995.

[41] S. Zheng, Composite optical fibreslab waveguides: a scalar, coupled-mode analysis with vector corrections, *Opt. Quant. Electro.* **29** (6) (1997) 671–682.

[42] M.K. Smit, Integrated optics in silicon-based aluminum oxide, PhD thesis, Delft University of Technology, Delft, The Netherlands, 1991.

[43] R.G. Hunsperger, *Integrated Optics*, Springer, Berlin, ISBN 3-540-59481-7, 1995.

[44] T. Tamir, (Ed.), *Integrated Optics*, Springer, Berlin, ISBN 0-3870-9673-6, 1979.

[45] P. Rai-Choudhury, *MEMS and MOEMS Technology and Applications*, Vol. **PM85**, SPIE Press Monograph, Washington, ISBN-0-8194-3716-6, 2000.

[46] Y.-C. Kim, H.-H. Park, J.S. Chun, W.-J. Lee, Compositional and structural analysis of aluminum oxide films prepared by plasma-enhanced chemical vapor deposition, *Thin Solid Films* **237** (1–2) (1994) 57–65.

[47] M.M. Aslan, N.A. Webster, C.L. Byard, M.B. Pereira, C.M. Hayes, R.S. Wiederkehr, Low-loss optical waveguides for the near ultra-violet and visible spectral regions with Al_2O_3 thin films from atomic layer deposition, *Thin Solid Films* **518** (17) (2010) 4935–4940.

[48] D.A. Roberts, R.R.A. Syms, A.S. Holmes, E.M. Yeatman, Dual numerical aperture confocal operation of a moving fibre bar-code reader, *Electron. Lett.* **35** (1999) 1656–1657.

[49] D.A. Roberts, R.R.A. Syms, 1d and 2d laser line scan generation using a fibre optic resonant scanner, *SPIE Proc.* **4075** (2000) 62–73.

[50] A. Duparre, S. Jakobs, Combination of surface characterization techniques for investigating optical thin film components, *Appl. Opt.* **35** (25) (1996) 5052–5058.

[51] D. Droste, J. Bille, A Hartmann-Shack wavefront detector realized with an ASIC, *IEEE J. Solid State Circuits* **37** (2002) 1–29.

[52] J.S. Dam, C.B. Pedersen, T. Dalgaard, P.E. Fabricius, P. Aruna, S. Andersson-Engels, Fibre-optic probe for noninvasive real-time determination of tissue optical properties at multiple wavelengths, *Appl. Opt.* **40** (7) (2001) 1155–1164.

[53] H. Nagata, M. Shiroishi, Y. Miyama, H. Mitsugi, N. Miyamoto, Evaluation of new UV-curable adhesive material for stable bonding between optical fibres and waveguide devices: Problems in device packaging, *Opt. Fiber Technol.* **1** (1995) 238–288.

[54] S.-Y. Lin, E. Chow, V. Hietala, P.R. Villeneuve, J.D. Joannopoulos, Experimental demonstration of guiding and bending of electromagnetic waves in a photonic crystal, *Science* **282** (1998) 274–276.

[55] Y.A. Vlasov, X.-Z. Bo, J.C. Sturm, D.J. Norris, On-chip natural assembly of silicon photonic bandgap crystals, *Nature* **414** (2001) 289–293.

[56] P.K. Tien, Integrated optics and thin-film technology, *Physica B + C* **89** (1977) 241–254.

[57] S.T. Chu, W.P. Huang, S.K. Chaudhuri, Simulation and analysis of waveguide based optical integrated circuits, *Comput. Phys. Commun.* **68** (1–3) (1991) 451–484.

CHAPTER

Chemical and Biological Sensors at Component and Device Level

6

CHAPTER CONTENTS

Developments in microchip-based chemical and biological sensors are closely driven by the requirements of products in the medical and safety devices industry. Specifically, the use of miniaturized devices at the so-called point-of-care, or point-of-action, is a strong driver for innovation in this field. Central laboratories have for many years offered a guarantee of access to state-of-the-art equipment and the highest quality of results for analytical (bio/medical) chemistry. However, logistic constraints and the large investments that are needed in infrastructural housing, as well as maintenance personnel, are extremely expensive in the diversified age of the 21st century.

Microfabrication for Industrial Applications. DOI: 10.1016/B978-0-8155-1582-1.00006-X

In addition, decentralized sensor networks can offer novel opportunities. Some problems cannot be solved well by central laboratories, because temporal and spatial information is an essential component of the solution. The development of decentralized sensor networks can tackle such problems, and have stimulated a new innovative wave in the measurement equipment sector. Microfabrication technology presents a solution, not just for decentralization, but also for the simplification of measurement protocols – to the extent that specialized operators are no longer required. One may summarize this trend as an *instant measure approach*. The pessimists amongst us may suggest that *Big Brother* is now not only watching us (enabled by a network of cheaply available and distributed video cameras accessible through the cost-efficiency of integrated electro-optical microchip manufacturing), but also punishing us immediately (for example, riding a bike through the warning signal of a traffic light, which may make you liable for an instant fine). The optimists on the other hand, consider the feeling of enhanced safety and fast response to act as quickly as possible to save a life. The same requirement for scanning open space has stimulated developments in the imaging of chemical and biological processes. This chapter gives an introduction to miniaturized chemical and biological sensors at the component and devices level, and also a sketch of a possible methodology for sensor development and testing.

6.1 APPLICATION FIELD

A great many chemical and biochemical sensor products already exist. They were first presented as part of the microelectronic miniaturization era of the early 1980s, when the research and development of electrochemical sensors gained much attention. In 1983, Japan organized the first conference dedicated to chemical microsensors. Micro-integrated chemical sensors are a niche market within the total sensor market, currently counting for approximately 10% of total sales. A large variety of different optical and electrical sensors have been explored in the context of microchip applications for chemical transduction in microreactor devices, due to scalability, in which a measurement can be obtained from very small quantities of gas or liquid.

We will discuss in more detail in the next sections how the need for monitoring is one of the keys for the successful roll-out of miniaturization in industry. For example, a microchip can be implemented to measure the composition of analytes in a chemical or biological sample during down-stream production processes. Such integrated analysis techniques allow product quality to be monitored early in the production line, e.g., using microchip chromatography. Microchip capillary electrophoresis also serves as a new chemical analysis method. Measurement of blood, plasma, serum, urine, droplets from the eye, saliva and spinal-cord fluid samples have already been demonstrated by on-chip measuring. Example applications include proteins, anti-epileptic drugs such as valproate in serum, and also measurements in whole blood, e.g., of anti-depressant pharmaceuticals such as lithium [1–4]. Chapter 7 will give details of the lithium measurement method, as part of a business case which illustrates the use of an integrated electroanalytical measurement system, in the point-of-care monitoring of lithium in patients with bipolar disorder. In the current chapter our discussion will concentrate on the component and device level of the sensor itself.

In many cases microchip techniques can be performed with a faster turnaround time than conventional analytical chemistry methods, but the larger scale methods are often more accurate. For point-of-care analysis, the user friendliness of the packaged device, and the robustness of the sensor design are as important as the analysis speed and accuracy.

To illustrate the environment in which such commercial sensors would operate, Figure 6.1 depicts a typical analytical chemistry laboratory environment (left) and the manual loading of a sample into a microplate for optical readout (right), as is often utilized for critical care patients. In the

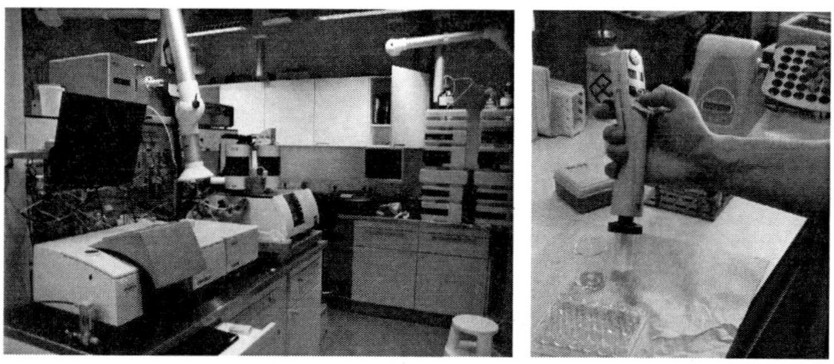

FIGURE 6.1

Typical chemical analytical laboratory facility with conventional apparatus (left) and example of manual sample loading to a microplate for optical read-out measurements in a biochemical-analysis laboratory (right), University of Twente. (Image courtesy: S. Schlautmann, University of Twente, 2011.)

pharmaceutical sciences these samples are mostly loaded automatically by dedicated robotic systems. In the operational environments of the chemical industry, flow-injection analysis (FIA) systems are often utilized, which decouple the harsh conditions of the product process line from the point-of-measurement by a bypass line housing the miniaturized chemical sensors. Timmer et al. present the μTAS approach for integrated microfabricated sensors, which can be utilized in a variety of industries. An example analyte is gaseous ammonia, which needs to be measured in environmental gas analysis as well as automotive, chemical and medical industrial applications [5]. Persaud et al. have presented an overview of "Solid state chemical sensors: Technologies and applications" in which they specifically addressed the requirements for sensor technology in the agricultural industry. The agricultural (bio)industry (intensive farming) needs reliable and affordable sensor technology for monitoring potential risks of environmental pollution, such as ammonia (NH_3), carbon dioxide (CO_2), hydrogen sulfide (H_2S) and methane (CH_4) [6].

Sensing, sampling and measurement technology as applied to real-time monitoring may require a totally different product configuration for each of the applications to which it is applied.

In the following sections we introduce working principles of sensors for chemical and biological measurands. Here, a biosensor is defined by the fact that it transducts the signal from the analyte by a biological recognition mechanism. We will briefly illustrate examples of sensing applications for measurands of both a chemical and biological nature, including low-molecular weight metabolites, metal ions, gases, anions and biologically active substances, for example, toxins, antibiotics and hormones. To observe and detect changes biologically, cells may be cultured and used to transduct, for example, the effects of pharmaceutics on living systems. The living cell is in this context a complex signal transduction system ("sensor") developing a positive or negative reaction upon exposure. As far as sensor integration technology is concerned, we are interested in a quantifiable change of the measurand, and a means to present these numbers to the user in an electronic format rather than an analog signal. The detailed evaluation of the technical context of such novel, microfabricated, sensor designs is, of course, very important, but so is the impact of widespread utilization of sensor technology throughout living environments and therefore on society.

6.2 SENSOR PRINCIPLES FOR THE COLLECTION OF (BIO)CHEMICAL INFORMATION

6.2.1 Optical Techniques

The increasing quality of signal response in analytical chemistry has benefitted from the development of ever better and more sophisticated photometric and fluorometric sensors. The miniaturization of such sensors has interesting applications in microreaction systems, portable systems for in-field and bed-side analysis applications, and in fermentation processes. The combination of different, discrete microsensors with fiber-optics or integrated microoptic components is an exciting research area. Optical set-ups that are sensitive to one specific, or multiple wavelengths may be used. The latter will produce light spectra, and thus can be utilized to deliver fingerprint results even of complex fluids, such as blood. For the determination of concentration changes of a specific substance in a background medium, single wavelength absorbance measurements (following the Lambert–Beer law) can be applied, where the intensity of the signal is proportional to the optical density of the medium. Specific markers, which change the optical properties of the measurand so as to produce a signal stronger or more selective. A marker can be introduced to interact with the incident light and produce either a fluorescent signal, or to enhance the optical absorbance of the compound at the selected wavelength. Figure 6.2 depicts a schematic and an example of a microscopic work place for the optical detection of signals as well as an image collected from a cell prepared with specific markers (also called staining of a cell or tissue material).

Biology laboratories would like to utilize the best optical microscope available, to study biological response curves in tissue and cells. So far, there has been little reason to reduce costs in the observation of bioanalytical processes, since we are just about to discover the potential of such a field of analysis to fulfill needs other than fundamental learning about cellular mechanisms. Once such analysis becomes important industrially, more automation will be required, and miniaturizing the optical imaging systems will offer additional advantages (see, for example, also new developments for compact optical detection in analytical systems-on-chip applications in Chapter 7, Section 7.7). Sensor systems that integrate optical components for (bio)chemical analysis are so far relatively rare, although hand-held

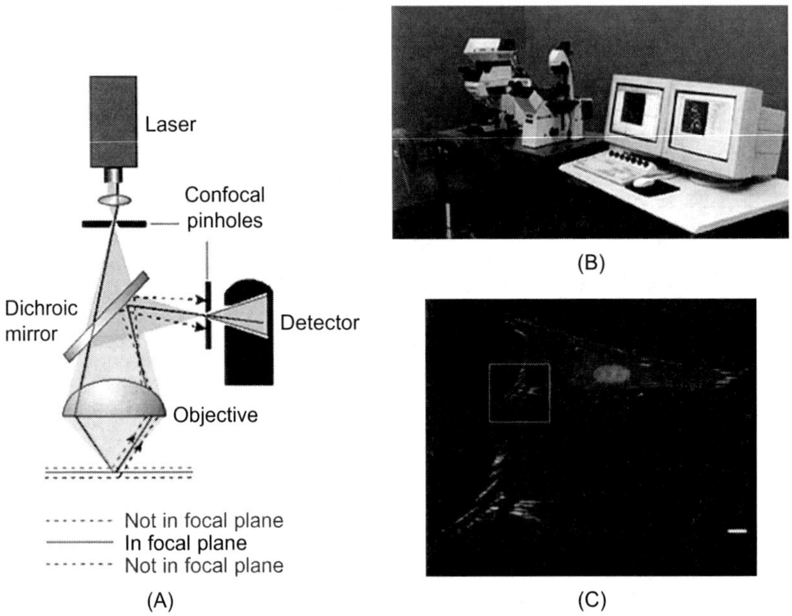

FIGURE 6.2

Optical microscopy: (A) Schematic of fluorescence imaging microscope and (B) inspection work place (original source: LEICA). (C) Example of immunofluorescence micrograph of osteoblast-like cells. (Image courtesy: E. Lamers, Radboud University Medical Center, 2009.)

photospectrometers delivering an electric readout are already available for certain puposes such as the color determination of teeth (shade detection) by a dentist. Another integrated optical measurement system is a miniaturized Mach–Zehnder interferometer, of which the fabrication process has been briefly described in Chapter 3.

Dasgupta et al. reviewed the use of light-emitting diode-based detectors within a planar flow-through cell, including a variety of application examples for integrated optics on chip [7]. One example of their application is the monitoring of water quality in the drinking water supply network. In these decentralized applications, optical systems are preferred over electrochemical sensors, since they can both generate and transmit the relevant data (an optical fiber telecommunication network is already available with a large area of

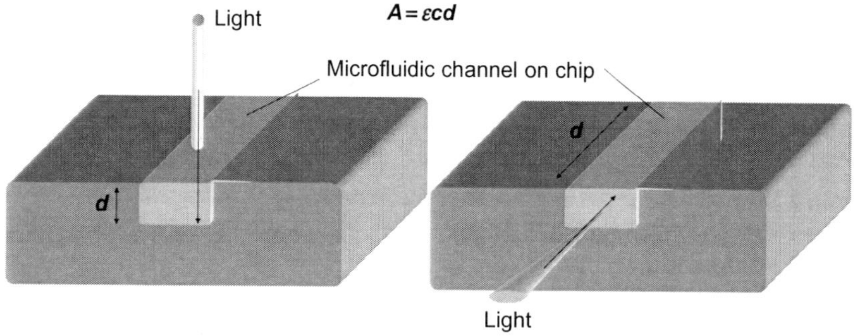

FIGURE 6.3

Principle of optical absorbance measurements, here d is the optical light path changing the reference signal by the concentration c in the microchannel.

coverage). In addition, they are insensitive to electromagnetic radiation, and thus are less prone to signal failure. Figure 6.3 illustrates how the signals of an on-chip fluidic network could be read, utilizing the method of absorbance measurement across the depth or the width of a microfluidic channel. This technique could be carried out on-chip using a microscope set-up, similar to the one shown in Figure 6.2B, fiber optics or the integration of planar waveguides into the chip. Methods to increase the optical path inside the chip by means of specially etched sidewalls of the flow channel that act as mirrors have been also described [8].

6.2.2 Electrochemical Techniques

Detection and quantification of chemical compounds is often carried out by electrical means, to reduce operator to operator variablity. In a hospital laboratory many samples must be measured rapidly for multiple parameters, and yet decisions have to be made with great care. The integration of electrochemical sensors is a suitable route for automation in this field. Chemical production plants are another potential example, where reading process parameters in real-time allows the utilization of such measurands for process loop control. Electrochemical microsensors can be used for process line monitoring across a range of simultaneous measurement points, bundling different levels of information into the decision-making process. Validating the quality of the signal, remains an on-going topic of debate if both the optical and the electrochemical principle of detection is possible. It may be more important to create sensors which produce an output that can be read directly than to make them robust. The signal validation problem can be solved, for example, by the installation of multi-sensor arrangements. The choice of sensor working principle and design of a sensor package depends very much on the specific field of application. There are many examples of such application fields, including pharmaceutics, medicine, analytical sciences, synthetic chemistry, biotechnology, material sciences and (bio)molecular engineering. The requirements of each for sensitivity, selectivity, portability, multivariant or single-use detection will all influence the choice of a suitable sensor principle, as well as its layout and constituent materials [9]. We will now give a brief introduction to two intensively investigated sensor principles, and then discuss some of their chemical and biological applications at component and device level.

Conductometry

If free ions are available in a solution, we call this solution an electrolyte. The positively and negatively charged ions are called cations and anions, repectively. An electrolytic solution behaves similarly to an electrical conductor, and its electrical properties follow Ohm's law: $I = U/R$, where I is the current, U is a voltage and R is the electrical resistance. R depends on the intrinsic properties of the conductor and its shape. This relationship is given as: $R = \rho \cdot l/A$, in which l is the length of the conductor, A is the cross-sectional area and ρ is the specific resitance. The resistance is also expressed as the conductance of the medium G, hence the name of the measurement principle. ρ is given in units of Ohm·m. The conductance G is given in units of S (Siemens), where 1S=1/Ohm.

We can imagine that an electrolyte could be sealed in a tube with a given length, and cross-sectional diameter A, and thus we can apply the same set of equations as for metallic conductors. To learn about the conductance of an electrolyte, measurements are made in a conductometric cell (container). The ratio l/A is an intrinsic parameter of the design of the conductometric cell and is called its resistive capacity C. This value may be calculated for a straight tube, but for complex geometries C is determined experimentally by using an electrolyte solution with a known specific conductance, for example, KCl solution. Once C is determined, solutions with unknown conductivity can be measured. The specific conductance is a function of temperature and so this must be measured carefully in applications of this sensor working principle.

The optimization of planar conductivity sensors suitable for lab-on-a-chip has been presented by Jacobs et al. [10].

Amperometry

Amperometric three-electrode measurement systems are state-of-the-art technology for the determination of toxins in the air, in which the gas diffuses into an electrochemical cell (electrolyte container with the electrodes). Planar microfabrication technology made it possible for systems to be batch fabricated at relatively low cost, and thus to be used as personalized meters for health and safety applications. The main advantage of microfabrication is the possibility to deposit different types of electrode materials acting as catalysts within a multi-sensor layout and reducing the electrode distance. Multi-layouts offer the potential for selective measurement of different specific compounds in a mixture and the reduction of the electrode distance enhances the sensor sensitivity [11]. The sensor working principle is based on the reaction of the electrochemical active compound at the working electrode, while new molecules diffuse towards the electrode and thus compensate the current flow. The electrical potential between the working electrode and the reference electrode is set and controlled by a potentiostat, while the current flowing between the working and the counter electrode provides the concentration-dependent signal.

The selection of the offered voltage at the working electrode depends on the measurand and if properly chosen, the measured current will be proportional to the amount of reduced compound. To control the mass transfer at the electrolyte–electrode interface, a diffusion-barrier is used by depositing a permeable material onto the electrode. These membrane-type materials should be deposited in as thin a layer as possible onto the electrode, to allow a fast sensor response and a sufficiently high sensor current. Membranes of several tens of micrometers may be easier to deposit than thin membranes, however the sensors will lose sensitivity. Developing a new microfabrication process is therefore always a trade-off between ease and robustness of manufacture and sensor performance. One of the best established and most investigated cases of amperometry is the electrochemical glucose sensor and its application in diabetes management [12].

Figure 6.4 shows a set up for amperometric sensing in a microfluidic channel structure using a conventional copper wire (left) and a variety of planar microfabricated electrodes (right) as studied by Cai et al. [13].

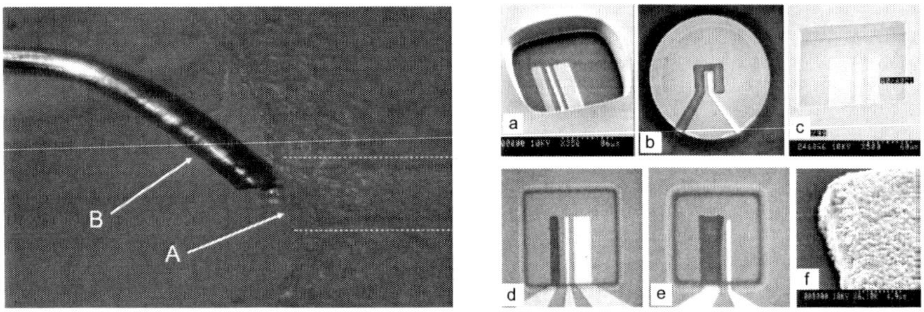

FIGURE 6.4

Left: principle of amperometric detection using a thin copper wire inserted into the microchip reservoir at the end face of the microchannel. (Image courtesy: E. X. Vrouwe, University of Twente, 2002.) Right: (a–e) different designs for microfabricated electrodes and (f) detail of electrodeposited electrode. (Reproduced with permission from reference [13]. Copyright 2000 John Wiley & Sons, Inc.)

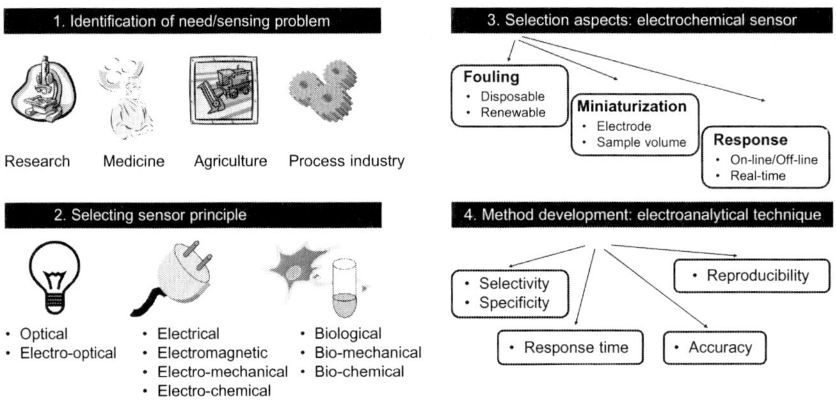

FIGURE 6.5

Overview of sensor development methodology.

6.2.3 Methodology of Sensor Development

Prior to the introduction of a novel sensor into a real-world analytical process, the components have to be tested at component level and assembled. Often the interfacing of different components can generate additional challenges, one of which is the reduction of dead volumes in the sensor design. However, joining two different materials can also cause an unexpected failure, particularly if electrochemistry is utilized, e.g., eroding the integrated electrodes.

Figure 6.5 presents a road map for the selection and design of a sensor principle and particular aspects to consider for electrochemical sensors. The design methodology for electrochemical sensors has been discussed previously, for example by Brett [14].

6.3 INTEGRATED chemFET DEVICE: CASE STUDY OF A SEMICONDUCTOR-BASED pH SENSOR DEVELOPMENT

pH is one of the most important parameters affecting chemical and biological reactions. As mentioned previously, it is common practice in process control and quality assurance to collect the process-related parameters downstream and use them to run (bio)chemical production processes safely and efficiently. The accurate measurement of a universal parameter such as pH or temperature often gives a good first order indication if a process is proceeding correctly. A shift in pH potentially indicates that the product quality has shifted, too. For production control, it is not always necessary to know what precise changes are occurring in product properties, although these may be measured during process optimization. Fluctuations in the environmental temperature at the measurement point may lead to error in reading the pH value and we need to look into the magnitude of these errors carefully depending on the customer's needs. The most established instrument for measuring pH is the pH-glass electrode, the measurement signal of which is a low voltage that is directly proportional to the pH value. For practical applications knowledge of pH sensing by conventional techniques is still useful; in this context we will consider how a change in pH influences the working of a semiconductor device.

Semiconductors can be used as chemical sensors in addition to their more widely-known applications [15, 16]. The so-called ion-sensitive field-effect transistor (ISFET) principle has been explored for the measurement of pH by Bergveld et al. [17–22] amongst others. The general ISFET design is based on the metal oxide semiconductor field effect transistor, in which the metal is replaced by an aqueous solution and a reference electrode, as depicted in Figure 6.6. The conductance between source and drain is a function of the electrical field perpendicular to the gate oxide surface. To measure pH, the ISFET's gatc oxide is exposed to the aqueous solution. If the pH in the solution changes, then the electronic signal will also change as a function of the state of ionization of the surface bound SiOH groups. For background on the working principles of semiconductor devices further reading is

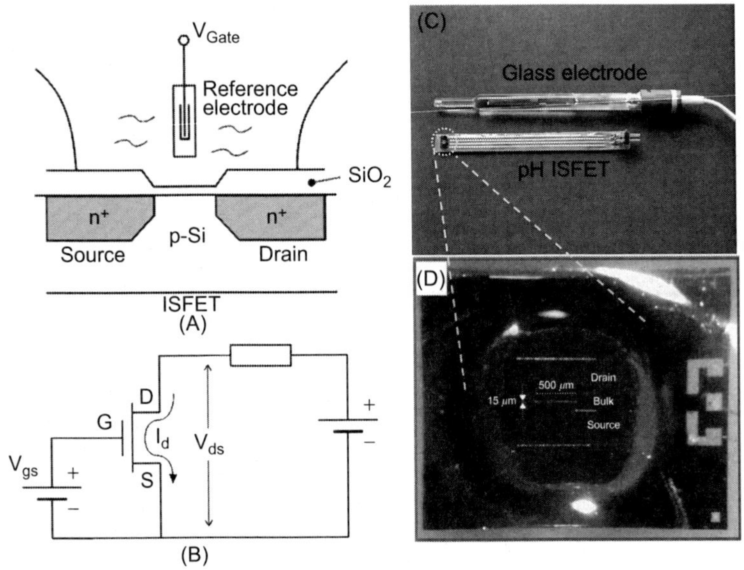

FIGURE 6.6

Microfabricated pH sensor. (A) Schematic of the working principle of an ion-selective field effect transistor (ISFET) and (B) ISFET electronic diagram. (Reproduced from reference [22].) (C) Realization of the packaged device compared to a standard glass hydrogen electrode and (D) detail of the silicon sensor element. (Images courtesy: P. Bergveld, University of Twente.)

encouraged, e.g., [23]. There is also considerable literature available on the ISFET, and at the University of Twente, The Netherlands, a number of PhD projects have focused on the development of this specific class of electrochemical sensors and the relevant PhD theses may be found through the University's library system, for example, de Rooij in 1978, Bousse in 1982 and van den Berg in 1988 [36–38].

6.4 INTEGRATED CLINICAL DIAGNOSTICS: A MEDICAL APPLICATION FOR ELECTROCHEMICAL SENSOR ARRAYS

Clinical diagnostics often relies on the measurement of a variety of blood parameters, e.g., the electrolyte balance or the blood gas analysis. The required analysis methods have often been developed in clinical chemistry.

Since the discovery of DNA and its importance to health patterns and prophylactic disease management, many molecular diagnostic assays on chip have been developed. The interaction sites are organized in a standardized x,y-matrix – so-called microarrays. These arrays may be read-out by optical means or by electrochemical sensors. Figure 6.7 depicts a schematic example of the assembly process of a DNA microarray.

6.4.1 From Microarray to Biochip Technology

Assays for method development in drug discovery, biochemistry and chemical sciences can also be performed on-chip. This is more efficient than conventional methods for bio-liquid analysis because of its high throughput and the much reduced amounts of solvents and reagents that are needed. This integrated microfluidic approach has already demonstrated remarkable results in a diveristy of life-science applications. Biolog Inc., for instance, followed this logical miniaturization roadmap with their Phenotype MicroArray™ technology. A BioChip design relies on the symbiosis between biology and physical instrumentation. In many applications biological phenomena of protein–protein and protein–cell interactions are investigated and monitored using microelectronic read-out systems that have been developed by microfabrication technology. DNA optical read-out has gained wide acceptance in applying fluorescence for gene expression profiling within the so-called microplate reader format. This approach has been used for the manipulation and analysis

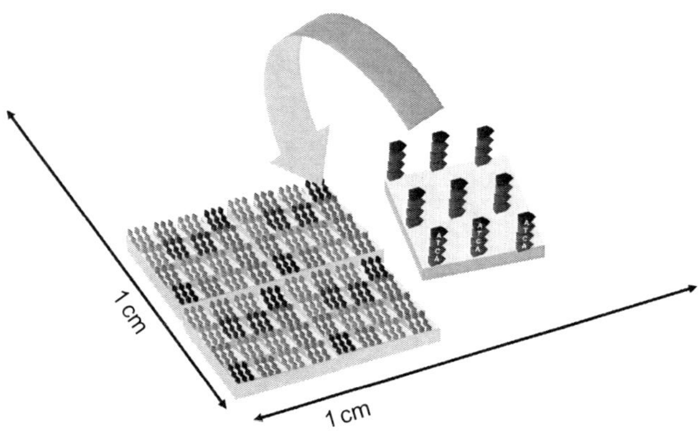

FIGURE 6.7

Schematic design of a DNA microarray.

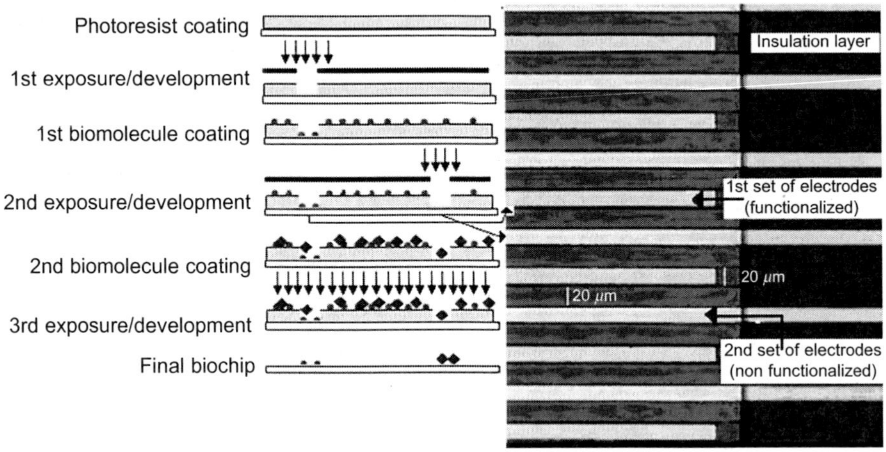

FIGURE 6.8

Micro-electrochemical biomolecular detection. Bio-photolithographic process for biomolecular patterning on the electrode array. On the right, gold interdigital electrode (IDE) array detail; each set of electrodes consists of 50 microelectrodes with a width of 20 μm and a distance between electrodes of 20 μm (Reproduced from reference [26]).

of thousands of different DNA sequences and proteins, either spotted or synthesized in-situ on a few cm^2 surface area [24, 25]. One example of a novel diagnostic sensor platform enabled by smart bio-photolithography and thin-film patterning is illustrated in Figure 6.8.

There is already a significant number of industrial players exploiting integrated label-free microarray approaches. Figure 6.9 illustrates an example of a bioelectrochemical integrated sensor array for advanced bio-assays. Such biochip designs may also employ flow-through systems with various types of biophysical detection methods. Infineon, for example, has developed such a type of array, based on their expertise in miniaturization for electronic systems, but in this instance using an optical read-out [27, 28].

6.4.2 Cell-Based Biosensor

In a new generation of transducers for the detection of biologically active substances, living cells can be utilized as the central unit [29, 30]. Recently, mammalian cell-based sensors have been investigated, as a tool for detecting pathogen and toxins in food, and for security applications with respect

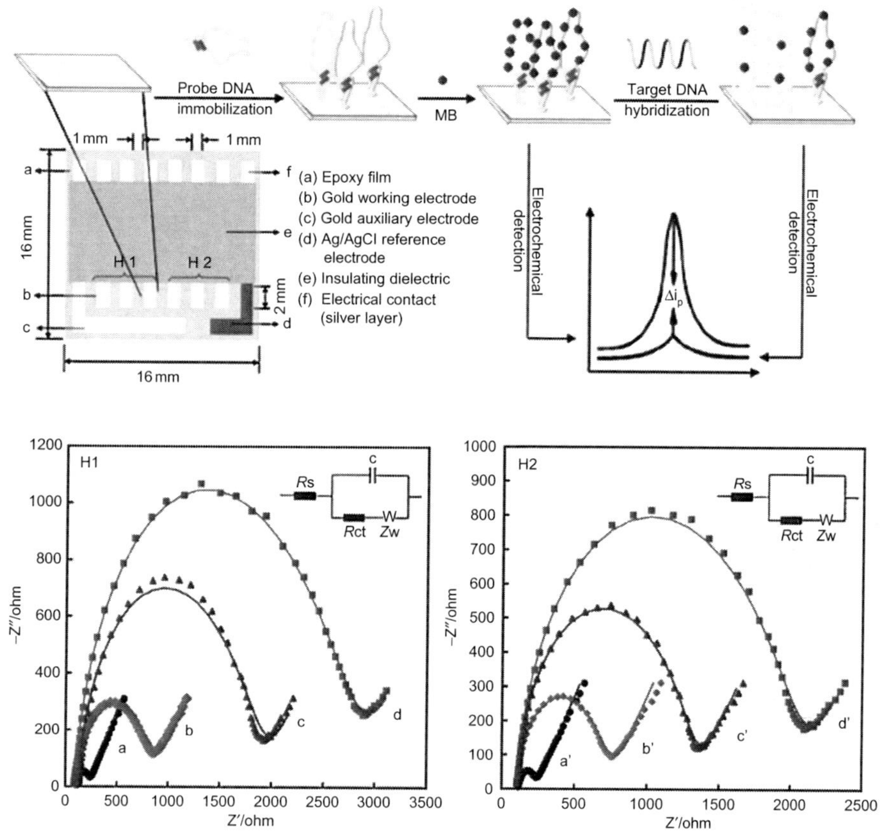

FIGURE 6.9

Top: schematic diagram of multi-electrode array and representation of biosensor array with fabrication steps and performance. Bottom: Faradic impedance spectra (FIS) of multi-electrode array in each immobilization step and binding step in 100 μL of 0.1 M PBS solution (pH 7.40) containing 10 mM $K_3Fe(CN)_6$-10 mM $K_4Fe(CN)_6$. H1: (a) Bare Au; (b) H1/Au; (c) H1/MCH/Au; H2: (a') bare Au; (b') H2/Au; (c') H2/MCH/Au; (d') HIV-2/H2/MCH/Au. Fitted data (solid line). The frequency was from 100 kHz to 0.1 Hz and the amplitude was 5.0 mV. The inset shows the equivalent circuit applied to fit the impedance spectroscopy. (Reproduced from reference [27].)

to defense strategics against chemical and biological attacks [31, 32]. A major challenge for the development of such systems is to keep the cell alive in a sealed microenvironment. Devices have been demonstrated using cell

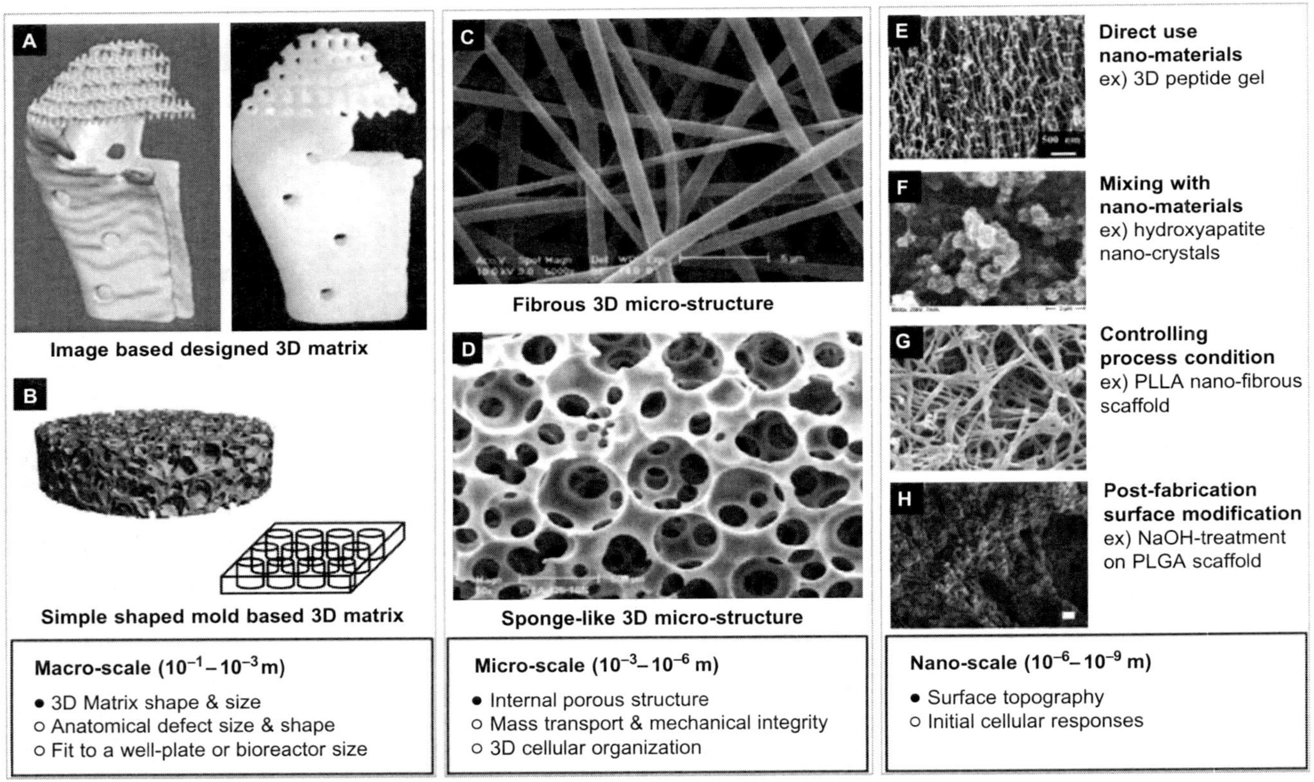

FIGURE 6.10

Design criteria of three-dimensional, multi-scale cell culture matrix. (Reproduced from reference [35]). The publisher for this copyrighted material is Mary Ann Liebert, Inc.

culture chambers made from polydimethylsiloxane (PDMS) and glass, while the electronic functions were realized in CMOS technology. Often in such cell-based biosensors the interaction between the cell and the electrochemical interface needs to be modified by cell-compatible molecules, such as molecules that enhance the adhesion of the cells to the sensor surfaces [33]. A variety of polymer grafting methods have been developed to coat the sensor's surface. Some of these polymers are combined with biological membrane proteins into an artificially designed surface layer. The selection of a specific surface modification also impacts the cellular response. Inappropriate mechanochemical transduction between surface and cell may even lead to cell death (apoptosis) or up- or down-regulation of gene expression may influence the observations of the pharmacological or toxicological response studies. Within stem-cell research, cells seeded on different types of surface structures may also differentiate differently, making the investigations even more complex.

Three-dimensional cellular networks have also been investigated. Cells and biomolecules may be deposited by principles similar to ink-jet printing. An example of this approach is given by Zhang et al., who demonstrated an industrial-scale, high-throughput, sandwich cell-based drug screening platform [34]. They applied micromachined Si_3N_4 sieves to enhance perfusion and maintain the three-dimensional morphology of the cells in culture over the long term. In addition to sensor technology, these techniques are also suitable for exploration in tissue regeneration processes. Figure 6.10 depicts an overview of different types of three-dimensional cell culture scaffolds that have typically been exploited in the artificial growth of tissues such as, for example, bone, cartilage, heart valves or skin [35].

6.5 CONCLUSIONS

Chemical processes and the intensification of (bio)chemical reactions rely on the optimization of dedicated technologies. These optimization studies can be carried out off-line, by measuring a set of feedback process parameters in a centralized laboratory. However, it is more efficient to measure the performance of a chemical reaction as early and as directly as possible during production processes. Appreciation of this fact led to the developments of the total analysis system (TAS). Originally, a TAS was simply a trolley containing all the equipment required for taking a reading from a specific production pipeline in the plant. Once the value of a specific parameter was known, a process optimization could take place immediately. With the ongoing requirement for cost reduction and even faster response times in industrial processes, microelectronic devices were used for the purpose of automation. The need to couple the read-outs into the (bio)chemical sensor systems quickly led to a new trend: microTAS (μTAS). This applied to chemical production plant optimization, as well as information retrieval concerning a variety of body functions that required rapid intervention.

At the component level, the electrochemical working principles can be integrated with widely established electronic functions and data-log systems. However, good design requires minimal dead volume and intelligent choice of materials.

Even more recent developments have seen multi-sampling of microarrays, by either optical or electrical sensors. The intensely high throughput demands of such fields as genomics and proteomics research have extended the boundaries of sensor developments to the point where the transduction processes of living cells are put to work in novel sensor designs. These novel platform technologies may be utilized in a variety of applications. The development of microsensors for universal parameters such as pH, conductivity and temperature are considered as milestones in chemical engineering, while the bio-recognition and interaction of molecules within specific biochip or microarray format gain importance for the understanding of biofunctions and their subsequent regulation in case of malfunctioning, e.g., during disease.

Unfortunately, the field of chemical and biochemical detection cannot be approached by a one-fits-all concept, and besides sensors for universal parameters such as temperature and pH, many distinguished sensor working principles and their working range performance are being investigated in the μTAS R&D community.

References

[1] C.L. Colyer, S.D. Mangru, D.J. Harrison, Microchip-based capillary electrophoresis of human serum proteins, *J. Chromatogr.* **A 781** (1–2) (1997) 271–276.

[2] C.L. Colyer, T. Tang, N. Chiem, D.J. Harrison, Clinical potential of microchip capillary electrophoresis systems, *Electrophoresis* **18** (10) (1997) 1733–1741.

[3] E. Ölvecká, M. Konīková, N. Grobuschek, D. Kaniansky, B. Stanislawski, Direct determination of valproate in serum by zone electrophoresis–isotachophoresis on a column-coupling chip, *J. Sep. Sci.* **26** (8) (2003) 693–700.

[4] E.X. Vrouwe, R. Luttge, A. van den Berg, Direct measurement of lithium in whole blood using microchip capillary electrophoresis with integrated conductivity detection, *Electrophoresis* **25** (10–11) (2004) 1660–1667.

[5] B. Timmer, W. Olthuis, A. Van Den Berg, Ammonia sensors and their applications – a review, *Sens. Actuators B Chem.* **107** (2) (2005) 666–677.

[6] K.C. Persaud, A. Flint, R.W. Sneath, Solid state chemical sensors: Technologies and applications, in: *EuroSime 2007: International Conference on Thermal, Mechanical and Multi-Physics Simulation Experiments in Microelectronics and Micro-Systems*, 2007, pp. 1–6, 16–18, April 2007.

[7] P.K. Dasgupta, I.-Y. Eom, K.J. Morris, J. Li, Light emitting diode-based detectors: Absorbance, fluorescence and spectroelectrochemical measurements in a planar flow-through cell, *Anal. Chim. Acta* **500** (1–2) (2003) 337–364.

[8] R.M. Tiggelaar, T.T. Veenstra, R.G.P. Sanders, J.G.E. Gardeniers, M.C. Elwenspoek, A. Van Den Berg, A light detection cell to be used in a micro analysis system for ammonia, *Talanta* **56** (2) (2002) 331–339.

[9] D.R. Thévenot, K. Toth, R.A. Durst, G.S. Wilson, Electrochemical biosensors: Recommended definitions and classification, *Biosens. Bioelectron.* **16** (1–2) (2001) 121–131.

[10] P. Jacobs, A. Varlan, W. Sansen, Design optimisation of planar electrolytic conductivity sensors, *Med. Biol. Eng. Comput.* **33** (6) (1995) 802–810.

[11] J. Kimura, T. Murakami, T. Kuriyama, I. Karube, An integrated multibiosensor for simultaneous amperometric and potentiometric measurement, *Sens. Actuators* **15** (4) (1988) 435–443.

[12] A. Heller, B. Feldman, Electrochemical glucose sensors and their applications in diabetes management, *Chem. Rev.* **108** (7) (2008) 2482–2505.

[13] X. Cai, A. Glidle, J.M. Cooper, Miniaturized electroanalytical sensor systems in micromachined structures, *Electroanalysis* **12** (9) (2000) 631–639.

[14] C.M.A. Brett, Electroanalytical techniques for the future: The challenges of miniaturization and of real-time measurements, *Electroanalysis* **11** (14) (1999) 1013–1016.

[15] C. Cané, I. Grácia, A. Merlos, Microtechnologies for pH ISFET chemical sensors, *Microelectronics J.* **28** (4 SPEC. ISS.) (1997) 389–405.

[16] A. Bratov, N. Abramova, A. Ipatov, Recent trends in potentiometric sensor arrays – a review, *Anal. Chim. Acta* **678** (2) (2010) 149–159.

[17] P. Bergveld, Ion sensitive field effect transistor, *Proc. Conf. Biocapt'75, International Conference on Biomed Transducers*, Paris, France, 3–7 November 1975, **1**, (1975), pp. 299–304.

[18] P. Bergveld, The operation of an ISFET as an electronic device, *Sens. Actuators* **1** (1) (1981) 17–29.

[19] P. Bergveld, B.H.v.d. Schoot, J.H.L. Onokiewicz, Development of a microprocessor-controlled coulometric system for stable pH control, *Anal. Chim. Acta* **151** (1983) 143–151.

[20] P. Bergveld, The development and application of FET-based biosensors, *Biosensors* **2** (1) (1986) 15–33.

[21] P. Bergveld, Future applications of ISFETs, *Sens. Actuators B Chem.* **4** (1–2) (1991) 125–133.

[22] P. Bergveld, Thirty years of ISFETOLOGY: What happened in the past 30 years and what may happen in the next 30 years, *Sens. Actuators B Chem.* **88** (1) (2003) 1–20.
[23] S.M. Sze, *Semiconductor Devices: Physics and Technology*, second ed., Wiley, Hoboken, NJ, 2001, ISBN 0-471-33372-7.
[24] A. Lagraulet, Current clinical and pharmaceutical applications of microarrays: From disease biomarkers discovery to automated diagnostics, *JALA* **15** (5) (2010) 405–413.
[25] I. Saaem, K.S. Ma, A.N. Marchi, T.H. LaBean, J. Tian, In situ synthesis of DNA microarray on functionalized cyclic olefin copolymer substrate, *ACS Appl. Mater. Interfaces* **2** (2) (2010) 491–497.
[26] M. Mir, S.K. Dondapati, M.V. Duarte, M. Chatzichristidi, K. Misiakos, P. Petrou, et al., Electrochemical biosensor microarray functionalized by means of biomolecule friendly photolithography, *Biosens. Bioelectron.* **25** (9) (2010) 2115–2121.
[27] D. Zhang, Y. Peng, H. Qi, Q. Gao, C. Zhang, Label-free electrochemical DNA biosensor array for simultaneous detection of the HIV-1 and HIV-2 oligonucleotides incorporating different hairpin-DNA probes and redox indicator, *Biosens. Bioelectron.* **25** (5) (2010) 1088–1094.
[28] N. Kessler, O. Ferraris, K. Palmer, W. Marsh, A. Steel, Use of the DNA flow-thru chip, a three-dimensional biochip, for typing and subtyping of influenza viruses, *J. Clin. Microbiol.* **42** (5) (2004) 2173–2185.
[29] L. Bousse, Whole cell biosensors, *Sens. Actuators B Chem.* **34** (1–3) (1996) 270–275.
[30] T.J. O'Shaughnessy, J.J. Pancrazio, Broadband detection of environmental neurotoxicants, *Anal. Chem.* **79** (23) (2007) 8838–8845.
[31] A. Bhunia, Mammalian cell-based sensors: A rapid screening tool for pathogens and toxins in food, *Food Eng. Ingredients* **35** (2010) 15–17.
[32] J. Janata, Role of analytical chemistry in defense strategies against chemical and biological attack, *Annu. Rev. Anal. Chem.* **2** (2009) 321–331.
[33] M. Ni, W.H. Tong, D. Choudhury, N.A.A. Rahim, C. Iliescu, H. Yu, Cell culture on MEMS platforms: A review, *Int. J. Mol. Sci.* **10** (12) (2009) 5411–5441.
[34] S. Zhang, W. Tong, B. Zheng, T.A.K. Susanto, L. Xia, C. Zhang, et al., A robust high-throughput sandwich cell-based drug screening platform, *Biomaterials* **32** (4) (2011) 1229–1241.
[35] J. Lee, M.J. Cuddihy, N.A. Kotov, Three-dimensional cell culture matrices: State of the art, *Tissue Eng. Part B Rev.* **14** (1) (2008) 61–86.
[36] N.F. de Rooij, The ISFET in electrochemisty: the influence of ionic compositions of solutions on the response of the ion-sensitive field effect transistor, PhD Thesis, University of Twente, Enschede, The Netherlands, 1978.
[37] L.J. Bousse, The chemical sensitivity of electrolyte/insulator/silicon structure: fundamentals of ISFET operation, PhD Thesis, University of Twente, Enschede, The Netherlands, 1982.
[38] A. van den Berg, Ion sensors based ISFET's with synthetic ionophores, PhD Thesis, University of Twente, Enschede, The Netherlands, 1988.

CHAPTER

Microfluidic Components, Devices and Integrated Lab-on-a-Chip Systems

7

CHAPTER CONTENTS

Defining the fabrication techniques and required features for mini-, micro- and nanofluidic devices, and clarifying their possibilities and drawbacks is becoming an important field of research and development. This chapter will summarize trends in microfluidic developments, and discuss microfluidic

Microfabrication for Industrial Applications. DOI: 10.1016/B978-0-8155-1582-1.00007-1

devices that will produce statistically meaningful results in chemical synthesis, chemical and pharmaceutical analysis, and biomedical diagnostics. It will also present a variety of systems that utilize the microfluidic-*inside* approach.

It is hoped that a layout database for miniaturized, fluidic sub-functional components, e.g., valves, pumps etc., will be established for the design of such systems allowing standardization of certain formats for pressurized inlets, seals, filling procedures and waste disposal, and enabling fast and effective communication with advanced microfabrication facilities for manufacturing on demand. This sort of database would allow chips to be customized during method development, and would speed up the engineering process from idea to product. Currently, trial and error plays a huge role in the design of microfluidic chip layouts by laboratory guided conceptual studies at an early stage, and dictates the implementation costs of a product. Although a conceptual approach is very important for the discovery of new phenomena, it hampers their commercial development. In application-driven research, these assembled sub-functions would ideally be tested similar to the modeling of electronic circuitry (e.g., utilizing dedicated modeling software). Intelligent design software has already been developed to some extent (e.g., marketed by Coventor, IntelliSense, etc.), but it needs to be further established in the innovation front-end process, and it is at present far from a universal standard.

Research programs can already benefit from externally defined fabrication lines for various types of processes and their product catalogs. For example, molding masters are ordered externally and are then used in relatively simple in-house replication techniques. A variety of commonly explored chip designs can also be ordered off the shelf (e.g., a capillary electrophoresis chip). An understanding of the fundamentals of pumping, valving, mixing and flow-monitoring are of prime interest during mini-, micro-, and nanofluidic research activities. This enables living organisms and cellular cultures to be investigated efficiently, but also a more efficient chemical screening can take place in miniature reactor designs (high throughput). Today, in chemistry and chemical engineering sciences many novel analytical methods and microreaction schemes are being developed using the microfluidic platform approach. Such an approach is safe and minimizes the need for resources during the optimization phases of chemical reactions for industrial use. This chapter focuses on some examples of microfabrication in the emerging markets for microfluidic applications.

7.1 APPLICATION FIELDS

Massively parallel screening methods that apply combinatorial (bio)chemistry in robotic-assisted fluid handling systems is one approach to handling extremely small samples. Microarray technology in medical diagnostics is another example of the implementation of experimental screening, which benefits tremendously from microfabrication techniques originating from the microelectronics industry. Expanding capabilities in the life sciences, which studies biological material, however, requires new solutions for chip manufacture. A variety of miniaturized, liquid-handling devices employ the principle of standardized microtiter plate formats. The miniature reaction vials are organized as an *x–y* matrix, which allows fast scanning across each of the reaction chambers. Flow-through devices allow continuous analysis on chip applying, for example, capillary electrophoresis (CE), chromatography or non-destructive optical techniques, all of which are integrated into one hand-held measurement system.

Continuing innovations in genomics and in analytical technology, provide pharmaceutical manufacturing and biotechnology businesses, such as brewing or food, it is hoped that these developments will benefit from new microfluidic methods. The concept of carrying out chemistry in confinement can allow the rapid development of new drugs and pharmaceutical treatments, as well as green and healthy food products.

Many innovations in these fields can be traced back to a basic phenomenon in fluid dynamics: the surface tension of incompressible fluids. A pressure is required to drive liquid through a capillary (microfluidic channel). The understanding of how different forces inside of capillaries interact such that they can be used effectively to transport compounds in these novel systems is an important area of research. Here, we will consider some working principles from a practical point of view, with the intention of being able to adapt a variety of operations known at the macroscale: pumping, dispensing, mixing, quenching, etc. Microfluidics can also be operated in combination with array strategies for feeding different samples to the chip, e.g., coupled with a mass-spectrometer. Many established conferences and peer-reviewed journals (*Analytical Chemistry, Electrophoresis, Sensors and Actuators, Journal of Microelectromechanical Systems, Transducers, μTAS*) have turned their attention to cell manipulation and analysis within artificial microenvironments. These arrangements allow well-defined spatial resolution for the observation of biological processes inside and in the extracellular matrix of living cells.

(Bio)chemical processes are investigated as part of the *in vitro* and *in vivo* testing of pharmaceutics, using animal models and volunteer human subjects. Pharmaceuticals may behave differently *in vivo* to what may be expected from *in vivo* studies; drugs may be unexpected effective against other diseases but can also cause undesired side effects. Knowledge of the overall (complex) behavior of a (bio) pharmaceutically active substance is difficult to generalize and statistically significant amounts of data have to be collected. Despite all the technological know-how, the risk of mistreating an individual is still high. Drugs can cause the agent of a disease to mutate, or cause undesirable changes in human genes (phenotype). Genotype and phenotype testing have become available for the diagnosis and evaluation of HIV drug resistance, and, in theory, resistance assays can be used to determine which drugs will be effective against an individual's strain of HIV, so that the appropriate treatment can be selected. There is still, however, much controversy about the accuracy of these diagnostic tools, and their usefulness as predictors of treatment success. Microfluidic-assisted systems for cell biology can help to elucidate these mechanisms.

Microfabricated devices can address cells, cell components and environmental features in the conglomerate of cultured and *in vivo* bio-environments, which will allow researchers to follow the dynamics of an individual cellular and molecular control mechanism in detail. This understanding of cell regulatory processes may be utilized to reconstruct even more complex bioprocesses including, for example, neuron circuitry, and bone and tissue repair mechanisms. These investigations may also identify an appropriate response to overcoming the risk of ill-treatment in case of pathogen mutation. Microfluidic devices can furthermore assist the biomedical sciences in general. This hypothesis concerns the controlled operation of moving, distributing and storing reagents and samples resulting in so-called cell culture chips with integrated analytical functions, e.g., utilizing the aforementioned integration concept of electrochemical and optical sensor technology. Miniaturized sensors are capable of interacting with cells at the same length scale, which may offer additional information.

The flow-through concept has several advantages over a batch-type approach, although both aim for high throughput solutions, massive parallelism and high reliability. Batch processes, using multiple pipettes, are prone to contamination, cross-interference, and are slower and less accurate in dynamic response when compared to the microfluidic approach, with its spatial discrete reaction and detection features. Additionally, titer plates are difficult to control due to evaporation, but if covered vials are used, fast robotic access will be limited. The polymerase chain reaction (PCR) "on chip" is a typical example of enhanced operation using a flow-through microfluidic device for finger-printing in cell biology. Microfluidics are also preferred for basic research in the life sciences, molecular diagnostics, pharmacogenomic profiling for personalized therapeutic planning, and the detection of food contaminants and bio-weapons. At present, titer plates not only serve as convenient sample or reagent storage trays, but additional functions are increasingly being integrated into the same platform. This will also allow rapid data correlation from 96 (384,.. etc.) analysis positions in a single run. The cavities of a titer plate can also serve as a reaction vessel or cell culture chamber, if they are nourished by a microfluidic network and allowing for cross-culture interactions. These advances in titer plate development, show the clear movement towards integrated microfluidics, which makes it a next logical step to pursue this trend and go for massively parallel flow-through microsystems for screening. Microfluidic on-chip processes are expected to improve the control of the reaction sites due to optimized heat and mass transfer. However, method development for *Lab-on-a-Chip* is still slow, and is particularly hampered by the high investment costs for customized chip technologies compared to the already well-established batch-based array-type assays in biology. Categorizing the various fluidic layouts into functional subunits, which can be assembled in a similar way to electronic design layout for standard components and better methods for *in silico* modeling, however, should make-up or this relatively slow progress in the near term future.

Advanced on-chip fluid electrokinetics will allow samples and other reagents to be prepared by more sensitive methods, and will also allow multiple measurements to be made from the same sample. Sample clean-up and enrichment prior to capillary electrophoresis has been studied earlier by conventional systems, for example, by Busher and co-workers [66]. These techniques, including selective extraction methods, are very suitable for microfabricated analytical systems, and it will be of great interest to develop integration strategies for microfluidics devices, often referred to in the MEMS community as micro-total analysis systems (μTAS), because of their original implementation as an analytical tool [1–3].

This chapter will consider the conceptual development of the lithium sensor as an example. I have been fascinated by the development of this device myself, probably to a greater extent than was strictly relevant. Innovation in industry is seeded by research projects. A good example of this is the LICETAS (Lithium Capillary Electrophoresis Total Analysis System), conducted in the BIOS-Lab-on-a-Chip group, headed by Prof. Dr. Ir. Albert van den Berg, at the University of Twente and financially supported by the Dutch Science and Technology Foundation (STW). Consideration of the LICETAS project may help us to answer an important question: How should governments and industry leaders make educated choices over which research should be funded, when all the suggested research projects are, of course, good ideas in the first place? LICETAS did not appear to be exceptional among the many excellent ideas presented by scientists, and was still some way from clinical application [4]. Recently the device has been further developed by Medimate BV, The Netherlands, and reported by Yager as a cutting-edge development for point-of-care diagnostics citing Floris et al. in *LabChip* [5, 6].

Clarifying the key differentiators versus the incremental progression of a scientific contribution may help in making educated decisions for funding research, either by government evaluators, or by industrial organizations looking for applications for their products. However, I can only let the section in this chapter speak for itself regarding whether the time and the money were well spent in turning the measurement of the conductivity of a liquid into a feasible concept for clinical point-of-care applications. In contrast to the many good books available in this field, I add my own personal experience of post-doctoral research. I hope to inspire readers to add their own knowledge to develop new micro- or nanofabricated products. Maybe there is no longer that much room at the bottom for miniaturization technology, however, focusing on the *More than Moore* principle in the applied sciences may successfully pave the way to an uncrowded business niche in industry [7, 8].

7.2 MICROFLUIDIC COMPONENTS

Generally, it is possible to bring a sensor to the measurement point, and simply expose the sensor to the environment of the measurand, but often this is not considered to be a practical solution. Particularly if the sample to be measured is blood, an off-line measurement is the method of choice. Also for many chemical engineering systems the decoupling of sample and production lines is often preferred, because the sample is crude and contains many more compounds than just the measurand. The sample needs to be worked-up, and sample preparation methods are needed if a measurement is to be conducted without interference from other compounds. Microfluidic components, and in particular a microvalve, could optimize sample preparation techniques, particularly when very small sample volumes are available, only. We therefore need to understand the control of transport phenomena of minute amounts of fluids (gases or liquids) when they are confined and moving from place to place. The efficiency of this transport is a particular concern in microfluidic systems. Most of the strategies using a variety of pumping mechanisms are inefficient at the small scale. On the other hand, at the very small length scale nature provides us with the phenomenon of diffusion as an efficient means of transport. Nevertheless, microscale operations, pumps, valves and mixers are still important structural elements for many microdevice operations. For details of the theoretical background of fluid mechanics, the study of continuum fluid dynamics equations and the laminar fluid flow patterns in confined spaces is important. With respect to the latter, I would like to refer to the physico-chemical background literature. An initial starting point for advanced reading is the compilation of publications from 2006 in the special issue on "Lab on a Chip" in *Nature* [9].

Microfluidic systems may be designed using any materials available in the microfabrication tool box: semiconductor materials, glasses, quartz, plastics, ceramics and even metals. With respect to valving and pumping, dynamic structural control is important. Examples do exist that use chemical interference on the valving or pumping structure, e.g., applying pH-induced volumetric changes in the case of elements fabricated from porous hydrogel materials, but mechanically rigid, structural mechanisms are still preferred in valve and pump designs. This is probably because they have a faster reaction time and more accurate mechanical placement. Prior to studying microfluidics it is strongly advised that the reader gains an overview of basic fluid mechanics at the macro- and mesoscale.

This book will introduce examples of microfabricated components for fluid handling, and the handling of the solutes in those fluids at the microscale in the next subsection. Some examples of non-moving structural elements that explore diffusion phenomena are also discussed. I anticipate that once some examples of the need for fluidic devices at this scale are illustrated, then the advanced study of transport and interface phenomena becomes much more exciting, and this study of illustrated practical cases will stimulate their industrial applicability.

A very good overview of the trend in micropumps, microvalves and micromixers has been presented within the context of the development of polymerase chain reaction (PCR) chips by Zhang et al. [10]. Specifically with respect to microvalves, Oh and Ahn presented a review article in 2006 [11]. Two years earlier, Laser and Santiago presented a review on micropumps [12]. Recently, these reviews were followed by an overview on pneumatic and hydraulic microactuators by De Volder and Reynaerts [13]. The latter discuss the importance of early MEMS research and development to microfluidics applications and illustrate a variety of designs in a comprehensive overview. Readers with a larger appetite for the design and application of fluid microactuation will doubtless highly appreciate the guidance on pneumatic and hydraulic microactuators given by De Volder and Raynaerts' review.

7.2.1 Passive Microvalves

Structures that regulate fluid transport inside the microfluidic network passively are structural elements that move by externally applied pressure and exert a force onto the mechanical element (spring-loaded diaphragm or cantilever-type structure). By means of this force the element that otherwise rests against the valve seal is lifted [14]. Obviously, these valves are of limited use in stand-alone miniaturized microfluidic applications, since a pressure must be applied that works against the pressure drop across the flow resistance of the flow conduit and the air pressure at the outlet. Based on this principle, it is obvious that strong external pumps are required for operation with these passive valves. One advantage of these structural elements is, however, that no energy is required when the valve is closed, which suggests a fail-safe operation.

As an alternative, two pressure-driven flows may be used in a splitter-type element, called a microfluidic flip-flop, that pushes the liquid stream to either one side or the other of the outlet channels, depending on a pressure pulse fed into the splitter by one of the symmetrically-shaped control ports. A pressure pulse in the side channel directs the liquid stream to the opposite wall, where it remains until the introduced gas has been vented, when a reverse pulse is applied. This type of set-up has been considered for microflow computing. Figure 7.1 gives schematics of these two passive structural valving elements. The microfluidic flip-flop may be operated in gradient mode with a sheath flow from the control ports for the generation of chemical dilution or washing steps on chip.

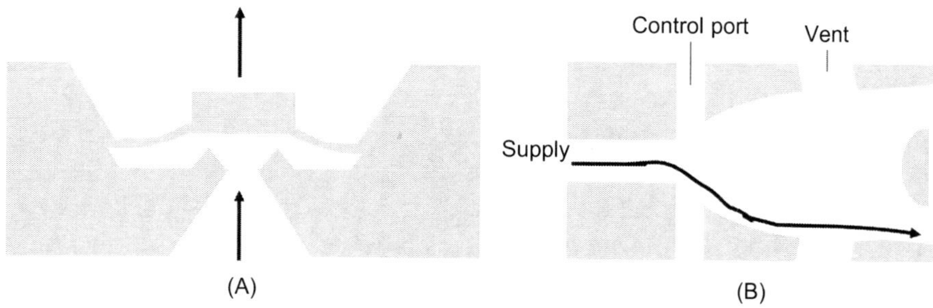

FIGURE 7.1

Passive microvalve structures. Diaphragm (A) in operating mode by increased inlet pressure. Microfluidic flip-flop structure (B) switching the flow direction between the two outlets by the pressure difference introduced by the normally symmetric control port flows.

Valving by the modification of material properties is also possible, for example, introducing a hydrophobic patch in a capillary can stop fluid transport. In this so-called stop valve the driving pressure must overcome the surface tension at the liquid meniscus. The threshold pressure depends on the hydrophobicity of the patch and its dimensions. In some Lab-on-a-Chip designs, a centrifugal force is applied to pass this type of barrier [15]. A similar effect may be obtained by an abrupt change in channel cross-section, which was demonstrated as one of the early, simple, valving principles in microfluidic devices [16].

7.2.2 Active Microvalves

Active microvalves operate normally "open", and close in a defined manner by the displacement of a moving element that regulates the flow in a microchannel. Active microvalves are considered to be those valves that have moving parts, or have material properties that are actuated by an external source of energy, for example, electrical, thermal, etc. Microvalve designs based on phase-changes in materials have been demonstrated, as have relatively simple designs of so-called thermal-viscous valves, which integrate a heating element with an orifice-type structure. These principles are often preferred to the aforementioned passive microvalve designs. For industrial applications, however, one needs to address the promise-requirement cycle of a selected working mechanism for a microvalve incorporated into a

Lab-on-a-Chip or microreactor design more carefully. Pumps and valves on a conventional scale may still be a better solution for many systems, and diaphragms and cantilever-type structural elements may be used to open or close the flow. The flow is either completely blocked, fully open or adapted by the remaining flow resistance of the valve. We described a set of microactuation principles in Chapter 5, which can also be applied to control microfluidic operations. A range of technical solutions have been presented in the scientific literature, or manufactured either at a research (functional demonstrator) or industrial scale (mass-manufactured devices). Due to their importance in microfluidic circuitry we will have a closer look at these structures in the following subsections.

Valving by Micromechanical Actuation

The micromechanical actuation of valves can be performed with the same microactuators as mechanical sensing applications, which we illustrated in Chapter 5. Combining the anisotropic etching of silicon and thin-film technology with electrical control circuitry and a means of a voltage applied to the actuator element; electrostatics will regulate the distance between the silicon boss structure and the microvalve seat. Adjusting the flow resistance at the valve orifice regulates subsequently the flow in the micro conduit. Figure 7.2 (left) depicts a schematic representation of this working mechanism. Alternatively, with the introduction of soft-MST utilizing PDMS a variety of interesting, pneumatically-actuated microfluidic valving structures

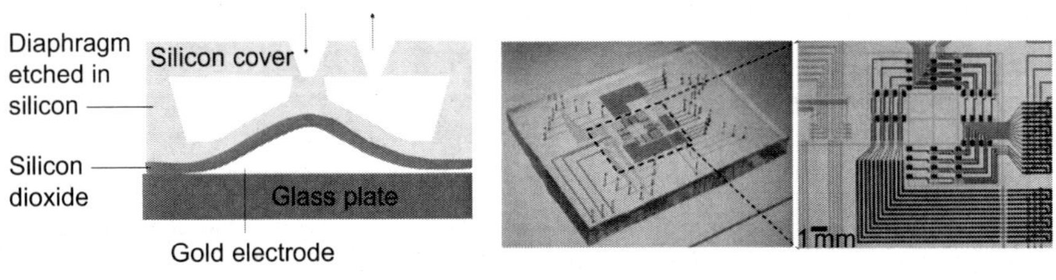

FIGURE 7.2

Micromechanical designs for valving. Left: schematic of an electrostatically actuated silicon valve enabled by anisotropic wet etching. Right: Optical micrographs depicting a PDMS-based microfluidic chip to generate and route droplets. The chip is fabricated by micromolding and lamination of two-layers of PDMS realizing multiplexed peristaltic pumps for high-density integrated microfluidics. (Reproduced from reference [17].)

can be made. An example is given by Cole et al. and is depicted at the right-hand side of Figure 7.2 [17]. Other actuator mechanisms, such as bimetallic, piezoelectric, or electromagnetic actuation can be coupled to diaphragm-type structures. Design modifications such as suspension bars can be included, which allow either to open up the flow through or in-plane of the wafer depending on the overall design of the microfluidic manifold in a specific application. Thermal expansion may be also used via changes of viscosity or temperature dependent phase-change materials. Gaseous compounds can be produced by electrochemical reactions in an enclosed reservoir and the increase in volume can be used to move a diaphragm. The latter mechanism was actually utilized in a design for an eye-pressure sensor.

Micromachined valves are also applied for the defined actuation of a Braille display, for example, utilizing specific conductive polymers in their designs [18]. A schematic design of a thermally actuated bimetallic microvalve is depicted in Figure 7.3. It was made by silicon etching, then a nickel structure was placed on top of the suspended silicon plate, by photolithography and electroplating. The thermally introduced mechanical stress in the bimetallic structure of approximately. 30 μm thick silicon and 30 μm thick nickel produces a

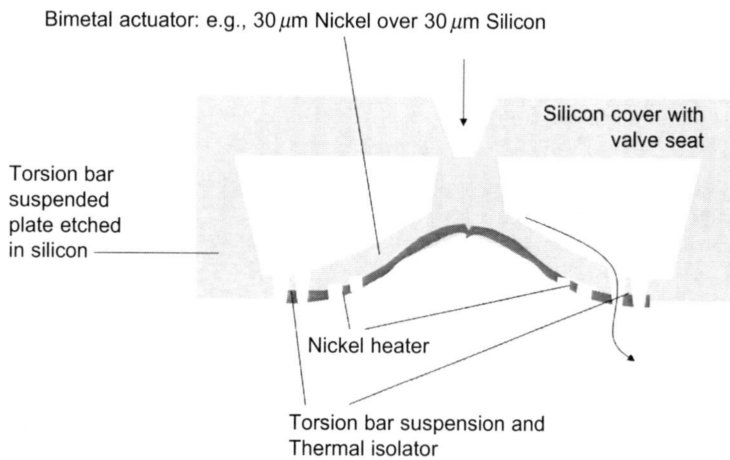

FIGURE 7.3

MEMS design for thermally actuated bimetal valve using electroplated Nickel as bimetal and heater. Schematic is based on silicon anisotropic etching as the principal microfabrication technology of the valve structure.

defined stroke between the boss and the valve seat. Surface micromachining and special materials may also be employed for micropumps and valves, as in the dual valve design described by Chen et al. [19].

7.3 CONTROLLED TRANSPORT BY DIFFUSION

Controlling chemical operations in a microfluidic chip may be achieved solely by diffusion. The laminar flow regime supports the operation of two streams merging from two different inlet channels into one channel, while the flow remains discretely side-by-side inside of this one channel. Solutes from one stream can now enter the other, and in the interface chemical reactions may occur. This exchange of solutes may be utilized for gradient dosing or extraction purposes. The exchange between the two streams is entirely limited by diffusion. Figure 7.4 demonstrates a simple H- and T-shaped microfluidic layout which can be produced by microfabrication in glass, or special plastics supporting optical transparency, and thus visualization of the events occurring in the device. An even simpler version may be designed from wicking paper. This was recently highlighted as a potential ultra-low cost solution for global public health care by Whitesides and Yager [20, 21]. By means of these types of structures a variety of applications have been demonstrated, some of which have also been commercialized as single-use medical diagnostic devices such as, for example, the H-Filter® and T-sensor® Access™ cards offered by Micronics, Inc., USA [22].

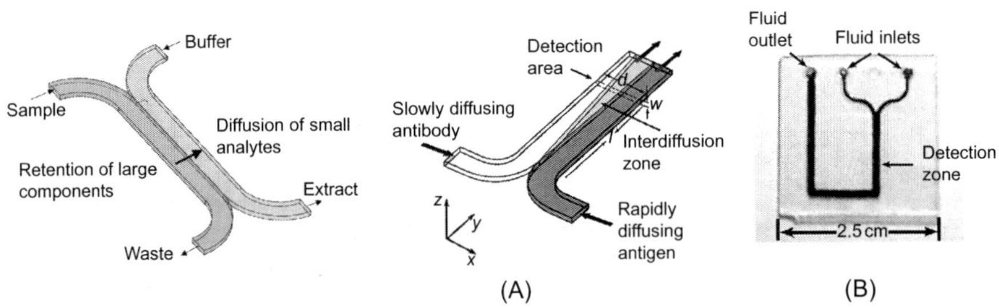

FIGURE 7.4

H-filter and T-sensor developed originally at the Yager Laboratory. (Left: reproduced from reference [22]; right (A, B): reproduced from reference [23].)

7.4 INTEGRATION FOR MICROFLUIDIC TRANSPORT, SENSING AND DISPENSING

In standard (bio)chemical handling, fluids containing analytes or reagents are transferred by means of a pipette. This may be an automatic pipette, with multiple tips loaded/deloaded in parallel, but single-tip pipettes are still in use, particularly in research environments. Using microfluidic transport for a similar operation in an experimental set-up on chip requires some level of component integration. The capillary effect, related to the above mentioned

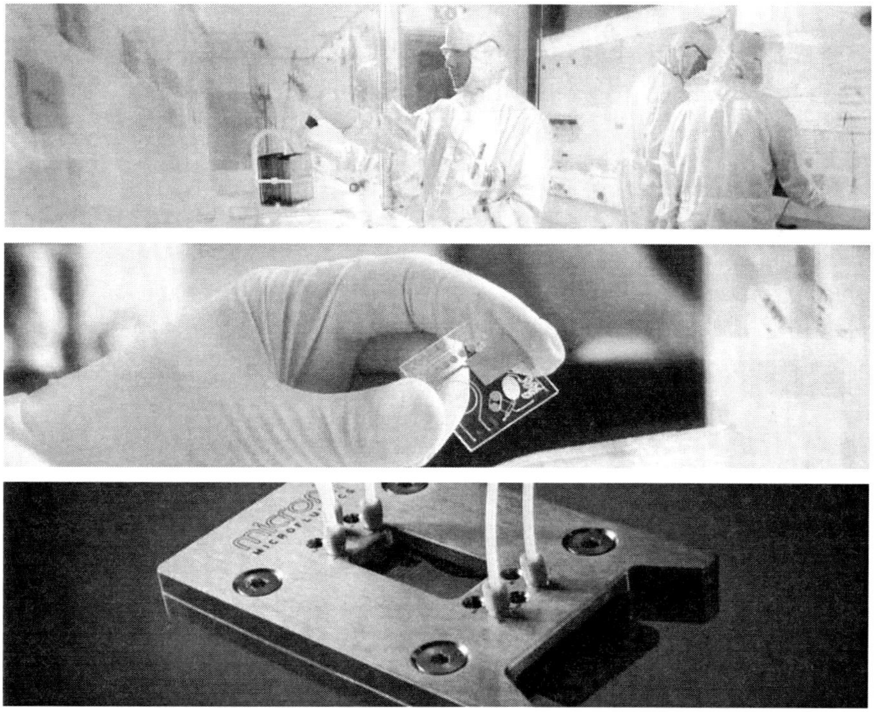

FIGURE 7.5

Product implementation from microfabrication to microchip testing by Micronit Microfluidics BV, Enschede, The Netherlands. Figures depict a view into Micronit's cleanroom facility (top), an example of an integrated glass chip with electrical contacts (middle) and Micronit's standardized chip holder for chip-to-world interfacing. (Image courtesy: Micronit Microfluidics BV, accessed online. [24].)

surface tension of the fluid and the size of the microchannel, could also be used. Hereby important parameters are the filling length of the channel h against the ambient air pressure at the end of the capillary proportional to the surface tension γ, the specific weight of the fluid w, the angle between the edge of the free fluid surface θ and the channel wall, and the radius a of the capillary tube. This basic concept is cleverly exploited in the diffusional set-up given in Figure 7.4. To eventually allow the lock-in handling of liquid dispensing after loading, a diverse range of integrated devices have been designed and run in an automatic, or at least semi-automatic fashion. A pumping action can be produced by other means besides simply pressurizing the inlet of a channel through a piston-driven, syringe pump. The application of an electrical potential across the capillary length will also result in a pumping action. This technique applies a field of up to a few 100 V/cm and is called electroosmotic fluid flow. The method appears to be easy, but it still requires some effort to miniaturize all the components, particularly the high-voltage power supply.

Leak-tide integration of electrodes has also demanded the attention of manufacturers in the field. Micronit Microfluidics BV, Enschede, The Netherlands, for example, has tackled the challenges of robust, integrated manufacturing in glass, and today offers a range of standard and customized microfluidic glass chips. Figure 7.5 illustrates these examples of product implementation in this novel, life-science oriented, microfabrication service industry.

7.5 LAB-ON-A-CHIP

The Lab-on-a-Chip originates from the initial development of miniaturized sensors and fluidic handling devices within one miniaturized package, the chip, so-called micro total analysis systems (μTAS) made up the first. A much broader implementation of these techniques quickly followed. There was considerable impetus for scientists already working on miniaturization to investigate means of measuring chemical reactions because of the large number of chemicals that are used in fabrication processes. While the latter not only attracted more and more scientists with a chemistry background to the research area of miniaturization technology, it also motivated the discovery of efficiency enhancement of chemical reactions performed in microfluidic channels.

Originally, Lab-on-a-Chip technology focused on analytical devices, and established such concepts as flow injection analysis (FIA) on chip. Throughout the last decade, however, we have experienced a revolution in this field in the life sciences. These have targeted a wide range of (bio)chemical reactions in cell cultures and clinical diagnostic point-of-care devices. The driver for this is the emergence of genetics as a medical field, where conventional analytical power is insufficient to sustain the required rate of experimental progress. So-called *sample-in-answer-out* capabilities are of considerable interest in life sciences [25–27].

Generally, micro- and nano-scale chemistry uses the increased surface-to-volume ratio resulting from the implementation of microfluidic devices. Combing μTAS and chemical synthesis research has led to the broader definition of this technical field, which is now known as Lab-on-a-Chip technology. The current prototypes translate the working principles of many standard laboratory analysis and synthesis techniques onto a chip card format (like a credit card). By fitting chemical reagents, test tubes, flasks and stirrers, pumps and valves onto a footprint hardly larger than a postage stamp has realized a paradigm shift in miniaturization. Figure 7.6 gives an introductory overview to the extensive developments of this novel field of technology.

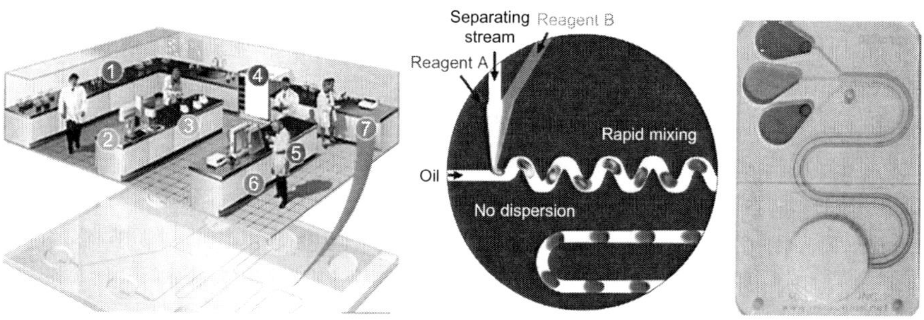

FIGURE 7.6

Laboratory processes on chip. Left: schematic illustration of translating key operations of conventional laboratories into chip formats. (Original source: Agilent, accessed online, 2005 [28].) Middle: a specific microfluidic set of operational steps on chip. (Reproduced from reference [29].) Right: an example of a commercial credit card-sized diagnostic device using a finger bellows that primes a wicking pad that then takes over the pumping action. (Reproduced by permission of The Royal Society of Chemistry [30].)

This on-going analytical revolution started with an extremely steep development curve in the 1980s, which approximately 30 years later, is reaching saturation of early curiosity-driven research. Although one cannot talk about a specific *killer application* this saturation clearly pays-off from a commercial point of view. The field of microfluidics controls the transport of gases and liquids inside the chip by a variety of structural components, as discussed in the previous section. Investigations of novel strategies for moving fluids in a channel network is described extensively in the literature. The initial developments in microfluidics were an off-spring of MEMS technology. These developments integrated valves and pumps for flow rate measurements, and mass sensors for highly defined gas supplies and safety equipment in the semiconductor industry. They also gave rise to pumps for controlled drug delivery for the chronically ill. Microfluidic circuits were also explored as far back as the 1970s in logic devices, because of their insensitivity to electromagnetic radiation.

Nowadays, microchemical reactions on a chip are popular as analytical systems for the same reasons as in analytical chemistry, and they are therefore branded as their own branch of Lab-on-Chip research and development, known as microreaction science and technology. Early on-line measurement systems were considered to be inferior to the possibilities of a well-equipped laboratory. Eventually, the introduction of flow-injection analysis, utilizing miniaturized systems was an economically driven choice. When it is more important in a bio(chemical) production line to have a rapid, crude measurement of product composition, miniaturization supported the rapid adoption of such μ-FIAs to be integrated into the production processes. The success of this supported the further validation of miniaturized sensor systems in field operations. This created a demand for the development of commercial microflow injection analysis (FIA) systems.

A FIA for (bio)chemical process lines is constructed as a mass flow controller for gases, in which a bypass line is utilized to carry out the measurement. One very successful example of chemical in-line measurement is the miniaturized gas chromatograph, marketed by Thermo-Fisher, who acquired the start-up company C2V BV. C2V developed this measurement system from concept to volume manufacturing.

Once the μ-FIA concept was introduced, researchers started to explore other application fields and to translate the micro total analysis approach to a variety of measurement challenges, including biology. The idea that biology could benefit from a system-in-a-card format is not all that obvious, since very sophisticated cell handling instrumentations already existed. Nevertheless, chips were introduced and tested, promoting the concept of automating cell manipulations at the level of a single cell. Microfluidic chip systems may actually be used to manipulate many thousands of single cells in parallel by integrating the capturing, triggering and analysis functions all into one system. Consideration had to be taken of the relevance to biology of taking measurements at the level of just one cell, since studies at the level of a population are usually necessary. Specifically, chips were considered as a means of learning more about cells in suspensions. Initial chip layouts utilized microfabrication to demonstrate miniaturized bio-handling systems similar to the operations that would normally be performed in an established fluorescent-activated cell sorter (FACS). FACS scanning the individual properties of the cells that previously had the opportunity for cell–cell interaction within the tissue population is a powerful tool for deriving statistically meaningful bio-information. Established non-chip formats allow a large number of cells to be counted by appearance (scatter, fluorescence from marker uptake, etc.), but without the possibility of tracing an individual cell in time throughout the experiment. If the single cell approach on chip is relevant biologically speaking then this design may be supported. However, the design of experiment with single cells remains a matter of further debate. The fact that *in-vitro* cultures of certain cell lines, in which cells are also isolated from their natural environment, are also a popular tool in biological experiments suggests that information collected from population, single-cell or tissue level can only model the *in-vivo* situation up to a certain level, and all the findings together may not produce a full understanding of real-life systems. We will not debate the various models for bio-medical experiments here, but will introduce the lab-on-a-chip concept and its technological realization by microfabrication as a potential approach for biological studies. Figure 7.7 shows the first microchip-based cytometer fabricated and tested by Micronics [31]. It clearly has been a milestone in the development of microfluidics for cell research.

The following subsections present an overview of microfluidics devices for particle and cell sorting, and miniaturized cell culturing systems, with a variety of integrated functionality. Micronics microfluidic card technology is based on laser micromachining, or stamping of foils and subsequent assembly by lamination.

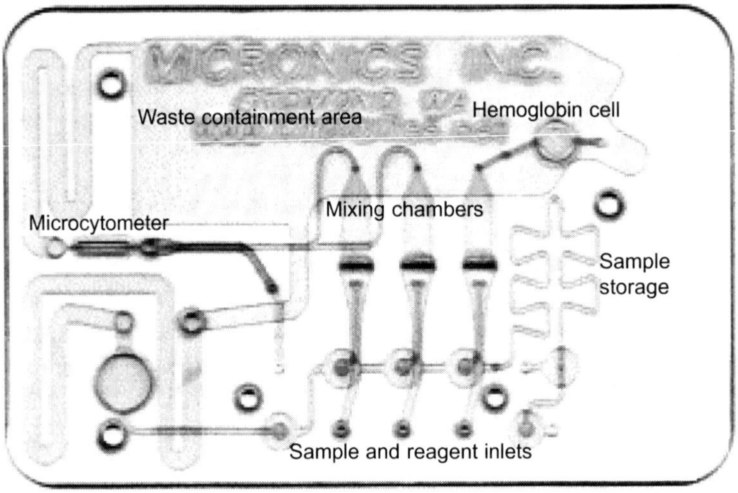

FIGURE 7.7

Micronics disposables are typically credit-card-sized, and most structural elements on these cards have dimensions ranging from about 100 μm to a few millimeters, use a low-cost rapid prototyping process that allows the design and testing of new microfluidic structures in 24 h or less. (Reproduced from reference [32].)

7.5.1 Miniaturized Particle and Cell Sorting Devices

Medical research and diagnostics depend on the careful identification of dysfunctional tissue or blood patterns. Conventional fluorescent-activated cell sorters (FACS) are used in clinical medicine, and basic biological and material sciences to isolate mutated cell types from healthy ones. A mutated cell type is then cultured, and undergoes multiple pathogen testing. To simplify such cell sorting procedures, miniaturized cell sorting and culturing devices have been demonstrated by Wolff et al. [33]. Current microfluidic device philosophy is particularly focused on finding strategies for cell sorting and trapping, and the subsequent manipulation and analysis of one cell at a time. Greater automation, and thus a faster turnaround is expected. Although the methods and protocols discussed above are clearly important for innovative approaches in the market, they will not necessarily mediate faster discovery. Improved medical treatments or insights into physiological parameters will not be possible without the careful observation of the cell in its natural context, i.e., together with neighboring cells and the extracellular matrix.

7.5.2 Cell Cultures and Fermentation Processes on Chip

Industrial bioreactors are macro environments where the culturing of a specific cell type is performed. Lab-on-a-Chip concepts are considered for yield-enhancing optimization studies of small-scale fermentation processes and for giving better control of the fermentation process. Investigations of optimized bioreaction processes within miniaturized cell reactors were introduced approximately 10 years ago [34].

7.6 DEVICE-TO-WORLD CONNECTIONS: THE MATAS CONCEPT

In Chapter 6 we discussed a set of sensors used in microfluidic devices. Microfluidics applications aim for hybrid integration in a similar way to the concept of combining different micromechanical and electronics sensors into one measurement system. The implementation of different devices and components into one platform is often referred to as the Lab-on-a-Chip concept, illustrated in Figure 7.6. Today, such systems have found their way from universities into chemical analysis, separation technology and synthesis (microreaction technology). The health-care market demands faster and faster response on differentiating disease patterns, and therefore a biolab infrastructure now demands much more flexible processes and rapid updates of protocols for faster screening.

Although conventional mechatronic robotics still competes with the revolutionizing possibilities of microfluidics, the latter is now widely recognized in the life sciences and industrial (fine-chemical) process control markets. The benefits of miniaturization in fluid handling environments are the reduced consumption of reagents, shorter temperature cycling times, faster mixing and a high degree of automation. These factors can result in a higher throughput, and a reduction in costs compared to conventional methods. One unique technology that is suitable to deliver a flexible and cost-effective realization of a Lab-on-a-Chip system is the MATAS (Modular Assembly for (micro) Total Analysis Systems), produced by LioniX BV, Enschede, The Netherlands [35].

MATAS is a highly standardized, modular technology for the design and fabrication of Lab-on-a-Chip systems-in-a-package. The technology enables the hybrid integration of microfluidic components, micro-optical devices and control electronics into one platform. This modular microfluidics kit has been selected for on-line measurement of pH and conductivity, and merges precision engineering with silicon micromachining. Figure 7.8 shows different components and devices of the MATAS.

Systems made up of these types of modules have so far been used as microreactors, sensors for pH, conductivity and ion concentrations, CE-chips, electrodes, flow sensors, micropumps, optical detectors, filters, valves and pressure sensors. All these different modules can be tested individually and exchanged just as quickly as ICs and resistors on a printed circuit board (PCB). The MATAS concept, originally developed by University of Twente's spin-off 3T, of which the microfluidic division was acquired by LioniX,

FIGURE 7.8

Different views of the components of the MATAS. The modular assembly PCB platform and components (top, left), detail of electronic interconnections (top, right), fluidic tubing (bottom, left) and silicon etched optical detection cell (bottom, right). MATAS may be utilized as a (micro-) Total Analysis and/or Lab-on-a-Chip system-in-a-package. (Image courtesy: LioniX BV [35].)

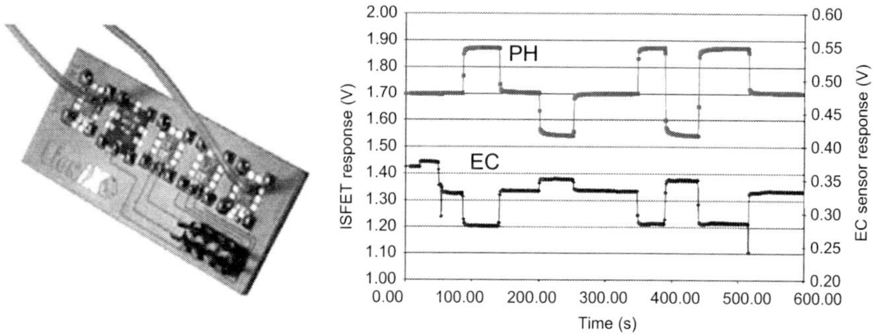

FIGURE 7.9

MATAS (left) for hybrid-integrated μTAS applications, here, ammonia detection. The traces (right) illustrate the measurement of pH and concentration changes in the analyte by integrated electrochemical sensor technology. (Image courtesy: LioniX BV, The Netherlands [35].)

allows the user to integrate hybrids of different components into one customized package. Modules can include optics-based sensors, actuators and microelectronics which are already manufactured in large-scale production lines, as well as structures produced by speciality foundries. Figure 7.9 depicts a fully assembled μTAS for the detection of pH and ammonia.

7.7 FROM THE LAB BENCH TO INDUSTRY: MICROCHIP CAPILLARY ELECTROPHORESIS

Capillary electrophoresis (CE) is a well established separation method in analytical chemistry. At the conventional scale small bore tubes must be used, in order to quickly dissipate the heat that is generated by the current flowing through the capillary at the applied high electrical field strength (several tens of V/cm). Under these conditions, the ions within the sample will separate, since they diffuse at different rates within the analytical medium. The separation length of a conventional system is approximately one meter because conventional fluid handling systems cannot produce the tightly defined picoliter-sized plugs that microfluidic implemented systems can. Figure 7.10 depicts an example of a conventionally operated CE system,

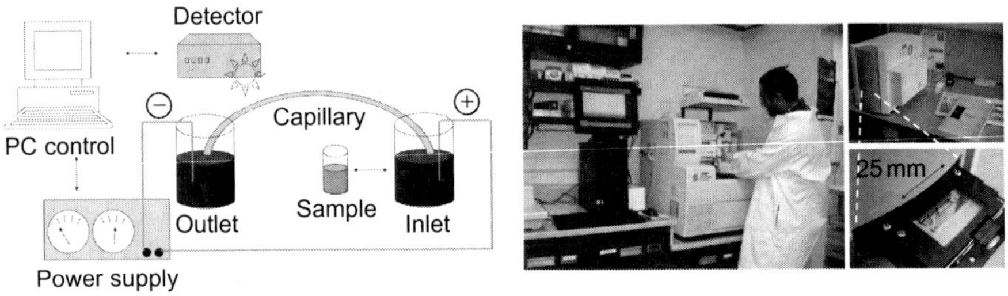

FIGURE 7.10

Left: schematic representation of the components of a conventional capillary electrophoresis (CE) set-up. Middle: established laboratory practice Beckman-CE work place for (bio)analytical chemistry. (Image accessed online, kindly provided by Dr. Stuart Le Grice, HIV Drug Resistance Program, National Cancer Institute-Frederick [36]). Right: desktop-integrated microchip CE system with optical UV-detection. (Image courtesy: E.X. Vrouwe, University of Twente, 2004.)

together with a chip-based CE apparatus with a total foot-print of a desktop computer, as developed by Shimadzu.

I was involved in LICETAS, a governmentally funded project, from 2001 to 2005 as a post-doctoral researcher together with an analytical chemist, Elwin Vrouwe, who at the time was working for his PhD. Dr. Vrouwe currently works as an industrial researcher at Micronit Microfluidics BV and I decided to continue my academic career at University of Twente. Nevertheless, I believe, it may be valuable to discuss some aspects of my involvement with the early efforts of the LICETAS innovation track in the context of this book. The project demonstrates an example of how a small seed-project of 25,000.00 Euros, granted for knowledge transfer to a spin-off company may make all the difference to industrial success.

LICETAS uses microchip electrophoresis with an integrated conductivity sensor, for the measurement of lithium in blood from a finger stick sample. This brings us to the set of subsections that describe the implementation of microfluidics devices at the point-of-care (PoC).

7.7.1 Is There a Need for a Microfluidic-Integrated System at the Doctor's Surgery?

Miniaturization technology is now capable of producing advanced analytical instrumentation that will fit within the palm of your hand. An answer to the above question may be: if technical progress has been made in this field, such that it may save a human life, we must therefore make every effort to explore its true potential. Once more we have to realize that it is the costs of implementation, and not the lack of good ideas or technical skills that delays the development chain from lab-bench to industry. As an example, the society's investment into the LICETAS research project and its initial utilization path was, roughly speaking, 500,000.00 Euros, which mainly paid the salaries of the researchers working on the project over a period of approximately 5 years (PhD candidate 4 years, postdoctorate 2 years, entrepreneur ca. 1 year). In society the following scenario exists: For one patient, who has to travel to a doctor's surgery, to get a blood sample measured by the centralized laboratory, the standard cost to the health-care provider is approximately 10 Euros per test. We can calculate that if this patient needs 10 tests per year, then the direct cost is just the 100 Euros per year, which will not justify the need for a microfluidic-integrated system at the doctor's surgery. However, the indirect costs involved in the whole procedure are enormous and their reduction will have also a significant positive impact on society as a whole and improve the care of the individual patient.

Although better health care is the primary driver of innovation, it is the economics that will determine whether or not these innovations are implemented. It is therefore important to analyze the business case, and carefully consider the indirect costs involved in the procedure of reading a patient's lithium level. Let us assume the following: 0.5 day traveling (home–lab–home), which means that the patient cannot work for at least half a day. With an average salary, this amount sums up to approximately 250 Euros (includes overhead for covering the time by a different member of staff). The patient may also need public transport or a taxi, at an average return ticket of 20 Euros. The total indirect costs for 10 tests per year therefore becomes 2700 Euros per patient per year.

In the Netherlands alone there are approximately 16,000 lithium cases. If we assume that we can halve the patient's traveling time (because in 5 of the 10 tests the patient will measure the lithum level at home or at work with the LICETAS), a total of 21.6 million Euros could be saved. Of course, this is just a rough business plan, and the development costs of the device also need to be covered, but surely this sum justifies the original 0.5 million Euros spent on the research project. So there is indeed a need for a point-of-care lithium measurement system, although not necessarily in the surgery, but in the palm of the patient's hand. Obviously, the full balance sheet of a company and a health-care provision facility are much more complex than the above example, but such simple financial overviews may give some guidelines on the decision-making process for the investment of public money.

7.7.2 The Technology Behind the Lithium Case

Lithium is a drug for the treatment of disordered brain functions. A psychiatrist needs to know the patient's metabolic profile on lithium turnover to decide on the appropriate dose. Ideally, measurements are performed at, or at least prior to visiting the doctor's surgery. This process may use a sample drawn from a finger stick only, or may need a whole blood sample taken from a venipuncture. Lithium is one of the most important mood stabilizers for treating bipolar mood disorders [37, 38], but one of its disadvantages is its narrow therapeutic range (0.4–1.2 mmol/L), which necessitates regular monitoring [39]. In the initial phase of the treatment the patient's lithium level is measured once or twice a week. Once the concentration has stabilized, the testing interval can be reduced to once every couple of months. Standard lithium monitoring has some limitations. Typically the samples are obtained from a venipuncture, which can be unpleasant for the patient. The whole procedure, including obtaining a sample, transporting it to the laboratory, sample pretreatment and the actual analysis, can take up to 45 minutes. Also, fluctuations in the lithium level remain undetected because of the low frequency of testing. In countries where there are few facilities where lithium can be determined, the lack of equipment or trained personnel may limit its prescription. A point-of-care test for lithium can solve all of these issues, offering enhanced treatment of patients. Nowadays the plasma or serum lithium level is often determined with ion-selective electrodes (ISEs) [40], which are also used in point-of-care tests to determine a range of parameters, including potassium and sodium from approximately 65 μl of whole blood [41]. However, a test for lithium using ISEs was not commercially available until 2004 [42]. This test measures the change in light absorbance by a porphyrin compound as it forms a complex with lithium. The sample is applied onto a cell separator strip, which is subsequently transferred to a cuvette filled with the reagent. These manual handling steps can make the test vulnerable to errors. Also, a considerable volume – 50 μl – is required for the test.

By contrast, a separation method such as capillary electrophoresis (CE) is not limited to a single analyte. Conventional CE is used for a wide range of substances, including lithium, serum proteins, amphetamines, β-agonists, organic acids and other substances [43–48]. Miniaturization offers the possibility of developing point-of-care tests. Not only is microchip CE faster than conventional systems, but integration of on-chip sample preparation can offer point-of-care testing [49, 50]. Microchip CE of inorganic ions, including lithium, in aqueous samples is relatively well established [51–55], but there is little information on handling and transfer of small sample volumes onto the chip.

The analysis of finger stick samples would be a great improvement over venous whole blood, and microneedles, which pierce the skin without inflicting pain, would be even more interesting [56, 57]. The latter have been used, for example, to sample interstitial fluid from the skin for glucose measurement [58]. In principle, these needles can be integrated relatively easily into microfabricated devices such as microchip CE, although the microchip needs to be optimized to work with the sample volumes, which are typically in the order of a few nanoliters. Initial results are promising, but so far only qualitative analysis of whole blood has been demonstrated [59].

Reverse iontophoresis offers an alternative method for painless sampling, and this method has also been used to extract lithium from the skin [60]. However, the analysis still has to be performed in the laboratory, and it can take up to an hour and a half to collect enough lithium for an accurate analysis. When immediate results are not required, reverse iontophoresis could be an interesting alternative to finger stick measurements as it has been demonstrated for the GlucoWatch® biographer. This commercially developed ultra-minimally invasive glucose monitor is based on a selective amperometric biosensor [61]. In the LICETAS project, both venous and finger stick samples were used to prove the principle of microchip analysis of lithium [62–64]. The aim was to transfer sample volumes of less than 10 μl to the chip by a disposable sample collection device, similar to a microneedle array. However, to reach this stage, a sample loading protocol and an analytical method had to be established. The CE separation conditions were optimized by enabling the direct analysis of samples by using on-chip conductivity detection. The results of this project demonstrated that a hand-held lithium point-of-care analyzer instrument had considerable potential. The following subsection will discuss different aspects of this novel point-of-care technology, which is currently being developed commercially by the University of Twente's MESA+ Institute for Nanotechnology spin-off, Medimate BV, Enschede, The Netherlands.

7.7.3 Microchip Capillary Electrophoresis Instrumentation

A computer-controlled, high voltage power supply, with four independently adjustable outputs was used to carry out microchip capillary electrophoresis measurements (CU 411, IBIS Technologies, Hengelo, The Netherlands). The power supply system also provided electrical current read out. A custom-made and commercially available conductivity detector was used for conductivity sensing, (Sprenkels Consultancy, Lelystad, The Netherlands). The detector signal was recorded with a data acquisition card (DAQCard 6036E, National Instruments, Austin, TX, USA), and an in-house written software package was used to control the power supply, and acquire and process the data, which included signal filtering and peak area calculations. Figure 7.11 depicts an overview of the components of the chip-holder, and also shows diagrams of the assembled microchip CE set-up, including the design. An image of a disposable blood-sample cup fitting the holder used in these experiments is also included in the figure.

7.7.4 Sample to Chip Interface

For the initial experiments, a glass capillary was used as a sample collector. Glass disposable micropipettes (0.5 mm inner diameter, 1.0 mm outer diameter, Drummond Scientific Company, Broomall, USA) were cut to a length of 1 cm. One end was sealed with a filter membrane (Millipore, mixed cellulose esters, 0.22 μm pore size) by heating the capillary in a gas flame and pressing it onto the membrane. The internal volume of the capillary, which defines the minimum amount of sample required, is 2.0 μl. A specific sample collector principle is tested for analysis of whole blood, by using a finger stick sample from a healthy volunteer. Capillary action draws the sample into the center channel, and up to the filter membrane, which, together with a water plug in the chip compartment, creates the sampling interface. Observation of the chip through a microscope shows that the cells do not go through the filter. The chip therefore stays clean, and can be used for multiple samples during initial reliability testing. This simple and easy to use sample collector helped to manage the experimental budget, because we could reuse the chip in pre-clinical trials of the system [64]. For subsequent testing, a disposable blood sample-cup could be used.

7.7.5 Samples

Five serum samples were received from the Medisch Spectrum Twente Hospital Group, Enschede, The Netherlands. The samples were obtained

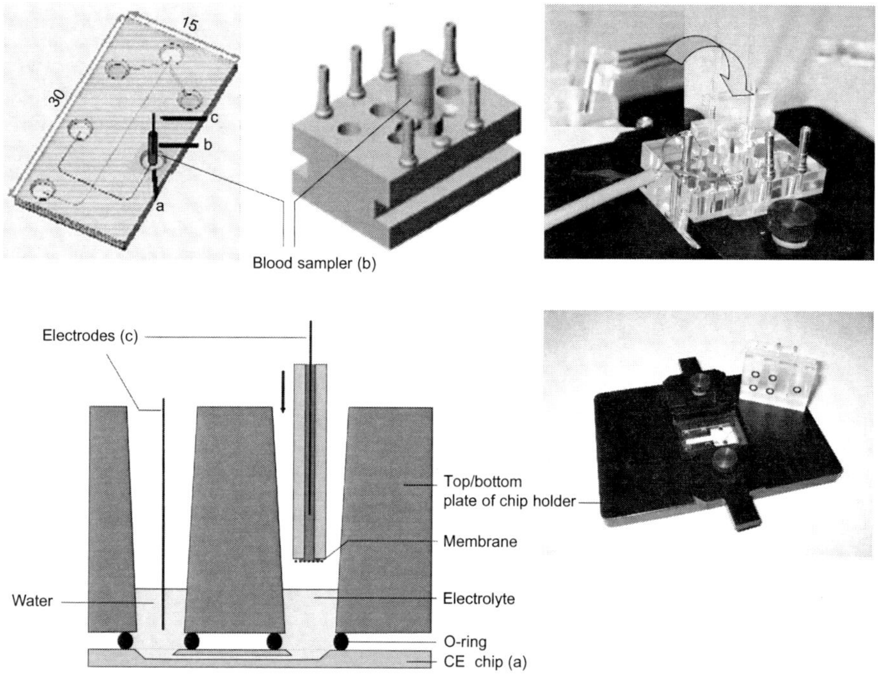

FIGURE 7.11

Top (from left to right): illustration of chip-layout (filled-grey circles indicate the contact pads of the integrated thin film electrodes at the end of the microchannel) and positioning of the blood sampler. Diagram of the chip holder top plate with the inserted blood sampler. Upper right corner: photograph of the assembled chip holder with an inset (upper left corner in the photograph) showing a close-up of the disposable and blood-filled sampler prior to insertion into the holder (transfer process marked with an arrow in the photograph). Bottom left: schematic of a cross-section view of the assembled chip/chip holder. Bottom right: de-assembled chip holder with a glass chip (30 mm × 15 mm) inside of the bottom plate. (Image courtesy: E.X. Vrouwe, University of Twente, 2005.)

from venipunctures of patients undergoing lithium therapy. The lithium concentrations were first determined in the hospital with an ion selective electrode (Cobas Integra 800, Roche, Basel, Switzerland). The lithium level was in the range of 0.49 to 0.90 mmol/L for all samples. The samples were subsequently analyzed by microchip CE on the same day. Additional samples were prepared as follows: whole blood was directly drawn into a sample collector after performing a finger stick on healthy volunteers, using disposable 21 gauge lancets with a puncture depth of 1.8 mm (Haemolance, HaeMedic AB, Munka Ljungby, Sweden). These samples did not contain any lithium, and therefore a second sample was collected in a sample vial by pipetting from the finger stick of the healthy volunteer, and spiked with 2 mmol/L lithium. Finally, a heparinized plasma sample was kindly provided by the Medisch Spectrum Twente Hospital Group from a patient on lithium therapy. This sample contained a lithium concentration of 0.62 mmol/L, as determined by the hospital.

7.7.6 Results and Conclusions from the LICETAS Project

A microchip assembly with a glass capillary interface as blood collector was successfully used for proof-of-prinicple in the laboratory to determine lithium in a whole blood sample from a healthy volunteer. A calibration sample is run first, followed by a whole blood sample from a healthy volunteer. Figure 7.12b shows the electropherogram containing sodium and magnesium in the typical ratio for whole blood. For quantitative measurements we used the lithium-spiked blood sample, as well as a heparinized patient plasma sample of known concentration. Both these samples were pipetted into a

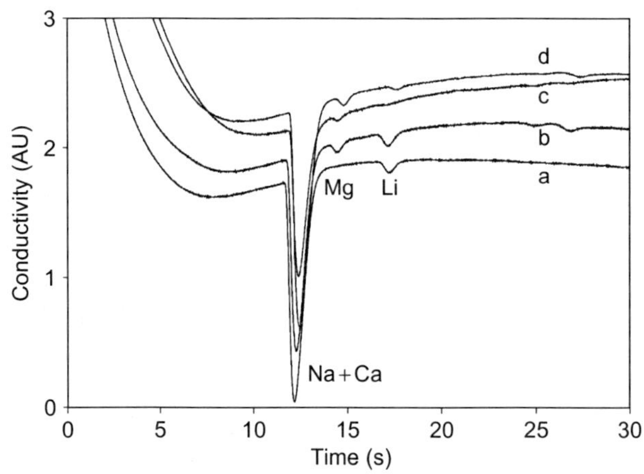

FIGURE 7.12

Electropherograms of (a) aqueous calibration mixture with 140 mM Na and 2 mM Li, (b) citrated whole blood spiked with 2 mmol/L Li, (c) whole blood without anticoagulant and (d) heparinized plasma from a patient on lithium therapy. Sampling via glass capillary. BGE 30 mmol/L ammonium acetate/acetic acid. (Image courtesy: E.X. Vrouwe, University of Twente, 2005).

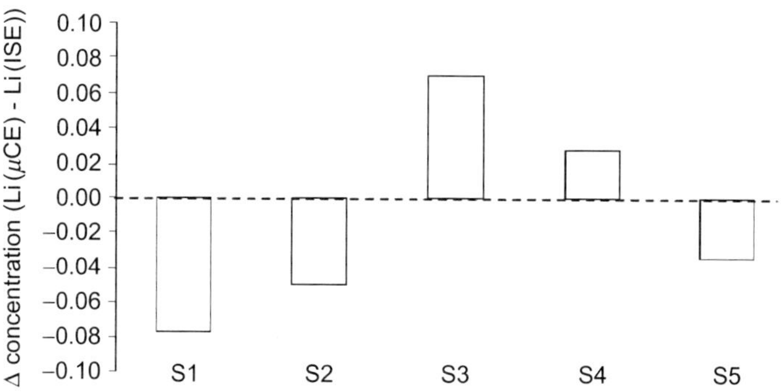

FIGURE 7.13

Comparison of the standard clinical values received by automated ion selective electrode (ISE) measurement with the values achieved by microchip CE method (μCE). Five patient blood samples (S1–S5) were tested. Bar diagram shows the difference of the two measurement values, which is within 10% of the clinical value.

sample collector for transfer to the chip. Figure 7.12c and d show the electropherograms of these two samples, respectively. The lithium peaks can be clearly identified from these data. The lithium concentration of the patient plasma sample was determined at 0.7 mmol/L (7% RSD, $n = 3$) using the calibration sample in Figure 7.12a, which is in good agreement with the hospital value of 0.62 mmol/L.

Figure 7.13 depicts correlation between the lithium determined by the CE microchip and an ion selective electrode at the hospital over five samples. Each point is the average of 6 measurements for the microchip system.

The microchip CE device is a versatile method for analyzing whole blood. Separations of sodium, magnesium, calcium and lithium are performed in less than 20 seconds without any off-system sample pretreatment. The disposable sample collectors presented here offer a convenient method for transferring a finger stick sample to the chip, and also serve to filter out potentially interfering blood cells. The chip itself does not come into direct contact with the sample, and can be used repeatedly for method validation. The sample stacking using a water plug worked reliably for three consecutive runs. Using sodium as an internal standard, clinically relevant lithium levels have been determined in serum samples, and the measurements are consistent with ion selective electrode results. The relatively high detection limit of 0.15 mmol/L and low precision need further optimization. This preliminary proof-of-principle gave an appropriate insight into the potential of this technique, and the results of this study were the first landmarks of the subsequent path for valorization, which has been initiated by the researchers on the governmentally funded project.

The knowledge gained through this project was successfully transferred from the University of Twenty's MESA+ Institute for Nanotechnology to its spin-off company Medimate BV. Based on additional seed-money from the *Valorisation Program* of the Dutch Science and Technology Foundation (STW), Mediate was able to take its first steps in using the scientific results, and translating the original concepts of the experimental set-up and the chip into a multi-reader version of a unique new product: a point-of-care solution for lithium measurements by the patient themselves. The product consists of a stand-alone reader (no external computer is required) and single-use sample cassettes with the chip. Both these units are depicted in Figure 7.14.

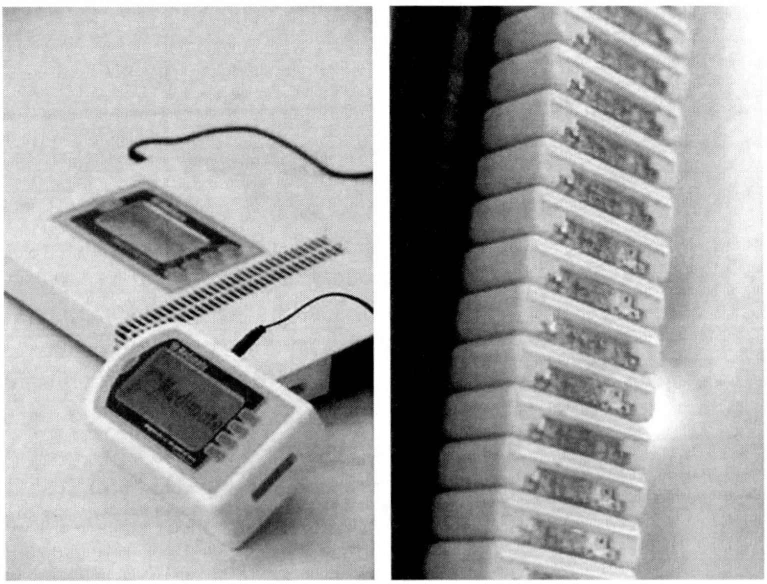

FIGURE 7.14

Multireader (left) and sample cassettes showing the mounted single-use chip for microchip capillary electrophoresis developed and manufactured by Medimate BV. (Image courtesy: Medimate BV [65].)

7.8 CONCLUSIONS

Original studies of the etching of micromechanical diaphragms for pressure sensors led first to the development of micromechanical actuators, and then to microvalves and pumps in microfluidics. Understanding diffusion and other transport phenomena, of course, is as important in confined environments as understanding the mechanics of pumping. The accurate fabrication of microfluidic channels in glass has opened up the possibility of studying micro- and nanofluidic behavior by advanced optical microscopy. Fluorescence and absorbance methods have been developed to determine the type and concentration of compounds in chemical mixtures. This plethora of novel opportunities has allowed the study of chemical reactions in-situ as well as the manipulation of delicate proteins or cells. Research on the optimization of soft-matter handling protocols is on-going in the field of Lab-on-a-Chip technology.

Studying and utilizing microfluidics has given rise to new applications. Although the microfluidic service industry is already well established, its full potential has yet to be explored. Possible avenues include chip layout services, provision of accessories and chip manufacture on demand. In this chapter, we discussed the potential for a lithium point-of-care microchip device. Estimation of the potential of these types of developments is not easy, and proof-of-principle studies (feasibility) can help to reduce the economic risks. Standards for this type of measurement protocol are yet to be formulated, so patenting is not a real indicator that the solution has an industrial applicability. However, it is applicability that is demanded by the European patent law actually for a patent to be granted. LICETAS was at a very early stage of development when it was transferred to a spin-off company, however, know-how was all the young start-up needed.

Work at the research stage showed the lithium peak appearing within the electropherogram drawn from a blood plasma sample as early as 2002; the method was not validated for its accuracy in a clinical setting. The results therefore were of generic interest to the scientific community, but not of importance to the medical diagnostics market because of the lack of a proof-of-concept for quantitative validation of the method. As soon as the quantitative reliability of the method could be demonstrated, the process of technology transfer to a company made sense. In due course, the required multidisciplinary know-how was collected by the newly formed spin-off company and turned into a novel product with an appropriate patent portfolio to commercially back-up the company's innovative technology.

Early stage technology transfer is a highly feasible approach to bring innovative solutions from the university to the market, but the LICETAS project operated at an extremely high risk to avoid its results being shelved as soon as the PhD thesis had been defended successfully. The fact that this did not happen was probably based on the right people being present in the right location, at the right time within one dedicated project team, and an appropriate funding agent (the Dutch Science and Technology Foundation, STW) sharing the one vision of bringing technical sciences to the marketplace.

References

[1] N. Blom, J.C. Fettinger, J. Koch, H. Ldi, A. Manz, H.M. Widmer, Implementing chemical sensors in industry: Novel approaches, *Sens. Actuators B Chem.* **5** (1–4) (1991) 75–78.

[2] R. Freaney, A. McShane, T.V. Keaveny, M. McKenna, K. Rabenstein, F.W. Scheller, et al., Novel instrumentation for real-time monitoring using miniaturized flow systems with integrated biosensors, *Ann. Clin. Biochem.* **34** (3) (1997) 291–302.

[3] M.U. Kopp, H.J. Crabtree, A. Manz, Developments in technology and applications of microsystems, *Curr. Opin. Chem. Biol.* **1** (3) (1997) 410–419.

[4] P. Wilding, S. Verpoorte, M. Allen Northrup, P. Yager, S. Quake, J. Landers, Microtechnology in the clinical laboratory: Will it solve analytical problems, and when will it make an impact? *Clin. Chem.* **56** (4) (2010) 508–514.

[5] P. Yager, Cutting edge: A prefilled, ready-to-use electrophoresis based lab-on-a-chip device for monitoring lithium in blood, *Lab Chip Miniaturisation Chem. Biol.* **10** (14) (2010) 1757–1757.

[6] A. Floris, S. Staal, S. Lenk, E. Staijen, D. Kohlheyer, J. Eijkel, et al., A prefilled, ready-to-use electrophoresis based lab-on-a-chip device for monitoring lithium in blood, *Lab Chip Miniaturisation Chem. Biol.* **10** (14) (2010) 1799–1806.

[7] K. Chakrabarty, Digital microfluidic biochips: A vision for functional diversity and more than Moore, in: *Proceedings IEEE Annual Symposium on VLSI*, ISVLSI 2010, 2010, pp. 3–4.

[8] K.D. Wise, Microelectronics in the "More than Moore" era, in: *Device Research Conference, Conference Digest*, DRC, 2010, pp. 3–4.

[9] J. Hogan, Lab on a chip: A little goes a long way, *Nature* **442** (2006) 351–352.

[10] C. Zhang, D. Xing, Y. Li, Micropumps, microvalves, and micromixers within PCR microfluidic chips: Advances and trends, *Biotechnol. Adv.* **25** (5) (2007) 483–514.

[11] K.W. Oh, C.H. Ahn, A review of microvalves, *J. Micromech. Microeng.* **16** (5) (2006), doi: 10.1088/0960–1317/16/SR01.

[12] D.J. Laser, J.G. Santiago, A review of micropumps, *J. Micromech. Microeng.* **14** (6) (2004), doi: 10.1088/0960–1317/14/6/R01.

[13] M. De Volder, D. Reynaerts, Pneumatic and hydraulic microactuators: A review, *J. Micromech. Microeng.* **20** (4) (2010), doi: 10.1088/0960–1317/17/7/00.
[14] D.C.S. Bien, S.J.N. Mitchell, H.S. Gamble, Fabrication and characterization of a micromachined passive valve, *J. Micromech. Microeng.* **13** (5) (2003) 557–562.
[15] M. Madou, J. Zoval, G. Jia, H. Kido, J. Kim, N. Kim, Lab on a CD, *Annu. Rev. Biomed. Eng.* **8** (2006) 601–628.
[16] P.F. Man, C.H. Mastrangelo, M.A. Burns, D.T. Burke, Microfabricated capillary-driven stop valve and sample injector, in: *Proceedings of the IEEE Micro Electro Mechanical Systems (MEMS)*, Univ of Michigan, Ann Arbor, United States, 1998, pp. 45–50.
[17] M.C. Cole, A.V. Desai, P.J.A. Kenis, Two-layer multiplexed peristaltic pumps for high-density integrated microfluidics, *Sens. Actuators B Chem.* **151** (2) (2011) 384–393.
[18] G.M. Spinks, G.G. Wallace, J. Ding, D. Zhou, B. Xi, J. Gillespie, Ionic Liquids and Polypyrrole Helix Tubes: Bringing the Electronic Braille Screen Closer to Reality, *Proceedings of SPIE – The International Society for Optical Engineering*, **5051**, 2003, pp. 372–380.
[19] P.-J. Chen, D.C. Rodger, E.M. Meng, M.S. Humayun, Y.-C. Tai, Surface-micromachined parylene dual valves for on-chip unpowered microflow regulation, *J. Microelectromech. Syst.* **16** (2) (2007) 223–231.
[20] A.W. Martinez, S.T. Phillips, G.M. Whitesides, E. Carrilho, Diagnostics for the developing world: Microfluidic paper-based analytical devices, *Anal. Chem.* **82** (1) (2010) 3–10.
[21] J.L. Osborn, B. Lutz, E. Fu, P. Kauffman, D.Y. Stevens, P. Yager, Microfluidics without pumps: Reinventing the t-sensor and h-filter in paper networks, *Lab Chip Miniaturisation Chem. Biol.* **10** (20) (2010) 2659–2665.
[22] P. Yager, T. Edwards, E. Fu, K. Helton, K. Nelson, M.R. Tam, et al., Microfluidic diagnostic technologies for global public health, *Nature* **442** (7101) (2006) 412–418.
[23] A. Hatch, A.E. Kamholz, K.R. Hawkins, M.S. Munson, E.A. Schilling, B.H. Weigl, et al., A rapid diffusion immunoassay in a t-sensor, *Nat. Biotechnol.* **19** (5) (2001) 461–465.
[24] Micronit Microfluidics BV. www.micronit.com (accessed 12.04.11).
[25] C.J. Easley, J.M. Karlinsey, J.M. Bienvenue, L.A. Legendre, M.G. Roper, S.H. Feldman, et al., A fully integrated microfluidic genetic analysis system with sample-in-answer-out capability, *Proc. Natl. Acad. Sci.* USA **103** (51) (2006) 19272–19277.
[26] R.G. Blazej, P. Kumaresan, R.A. Mathies, Microfabricated bioprocessor for integrated nanoliter-scale sanger DNA sequencing, *Proc. Natl. Acad. Sci.* USA **103** (19) (2006) 7240–7245.
[27] C.A. Hutchison III, DNA sequencing: Bench to bedside and beyond, *Nucleic Acids Res.* **35** (18) (2007) 6227–6237.
[28] Agilent Technologies. http://www.agilent.com (accessed 12.04.11).
[29] H. Song, J.D. Tice, R.F. Ismagilov, A microfluidic system for controlling reaction networks in time, *Angew. Chem. Int. Ed.* **42** (7) (2003) 768–772.
[30] B. Weigl, G. Domingo, P. LaBarre, J. Gerlach, Towards non- and minimally instrumented, microfluidics-based diagnostic devices, *Lab Chip Miniaturisation Chem. Biol.* **8** (12) (2008) 1999–2014.
[31] Micronics. http://www.micronics.net (accessed 12.04.11).

[32] B.H. Weigl, R.L. Bardell, C.R. Cabrera, Lab-on-a-chip for drug development, *Adv. Drug. Deliv. Rev.* **55** (3) (2003) 349–377.

[33] A. Wolff, I.R. Perch-Nielsen, U.D. Larsen, P. Friis, G. Goranovic, C.R. Poulsen, et al., Integrating advanced functionality in a microfabricated high-throughput fluorescent-activated cell sorter, *Lab Chip Miniaturisation Chem. Biol.* **3** (1) (2003) 22–27.

[34] D.A. Fitzgerald, Macro opportunities in microfluidics, *Scientist* **7** (2003).

[35] LioniX microfluidics. www.lionixbv.com (accessed 12.04.11).

[36] http://home.ncifcrf.gov/hivdrp/SLG_Fig11.jpg. http://home.ncifcrf.gov/hivdrp/SLG _Fig11 .jpg; http://home.ncifcrf.gov/hivdrp/Le_Grice_lab.html, 2011 (accessed 12.04.11).

[37] N.J. Birch, Inorganic pharmacology of lithium, *Chem. Rev.* **99** (9) (1999) 2659–2682.

[38] B. Müller-Oerlinghausen, A. Berghöfer, M. Bauer, Bipolar disorder, *The Lancet* **359** (9302) (2002) 241–247.

[39] A. Amdisen, Monitoring of lithium treatment through determination of lithium concentration, *Dan. Med. Bull.* **22** (7) (1975) 277–291.

[40] S. Linko, Automated ion-selective measurement of lithium in serum. A practical approach to result-level verification in a two-way method validation, *Accredit. Qual. Assur.* **6** (2001) 31–36.

[41] K.A. Erickson, P. Wilding, Evaluation of a novel point-of-care system, the i-STAT portable clinical analyzer, *Clin. Chem.* **39** (2) (1993) 283–287.

[42] W.M. Glazer, J.G. Sonnenberg, M.J. Reinstein, R.F. Akers, A novel, point-of-care test for lithium levels: Description and reliability, *J. Clin. Psychiatry* **65** (5) (2004) 652–655.

[43] X. Huang, M.J. Gordon, R.N. Zare, Quantitation of Li^+ in serum by capillary zone electrophoresis with an on-column conductivity detector, *J. Chromatogr. B Biomed. Sci. Appl.* **425** (2) (1988) 385–390.

[44] V. Dolnik, Capillary electrophoresis of proteins 2003–2005, *Electrophoresis* **27** (1) (2006) 126–141.

[45] G. Boatto, M.V. Faedda, A. Pau, B. Asproni, S. Menconi, R. Cerri, Determination of amphetamines in human whole blood by capillary electrophoresis with photodiode array detection, *J. Pharm. Biomed. Anal.* **29** (6) (2002) 1073–1080.

[46] T. Zhou, Q. Hu, H. Yu, Y. Fang, Separation and determination of β-agonists in serum by capillary zone electrophoresis with amperometric detection, *Anal. Chim. Acta* **441** (1) (2001) 23–28.

[47] J. Boden, K. Bächmann, Investigation of matrix effects in capillary zone electrophoresis, *J. Chromatogr. A* **734** (2) (1996) 319–330.

[48] R. Jabeen, D. Payne, J. Wiktorowicz, A. Mohammad, J. Petersen, Capillary electrophoresis and the clinical laboratory, *Electrophoresis* **27** (12) (2006) 2413–2438.

[49] J. Cheng, L.J. Kricka, E.L. Scheldon, P. Wilding, Sample preparation in microstructured devices, *Top Curr. Chem.* **194** (1998) 215–231.

[50] D.D. Cunningham, Fluidics and sample handling in clinical chemical analysis, *Anal. Chim. Acta* **429** (1) (2001) 1–18.

[51] R.M. Guijt, E. Baltussen, G.D. Van Steen, R.B.M. Schasfoort, S. Schlautmann, H.A.H. Billiet, et al., Fabrication of quartz microchips with optical slit and development of a linear imaging UV detector for microchip electrophoresis systems, *Electrophoresis* **22** (2) (2001) 230–234.

[52] M. Pumera, J. Wang, F. Opekar, I. Jelinek, J. Feldman, H. Löwe, et al., Contactless conductivity detector for microchip capillary electrophoresis, *Anal. Chem.* **74** (9) (2002) 1968–1971.

[53] J. Lichtenberg, N.F. de Rooij, E. Verpoorte, A microchip electrophoresis system with integrated in-plane electrodes for contactless conductivity detection, *Electrophoresis* **23** (21) (2002) 3769–3780.

[54] J. Tanyanyiwa, E.M. Abad-Villar, M.T. Fernndez-Abedul, A. Costa-Garca, W. Hoffmann, A.E. Guber, et al., High-voltage contactless conductivity-detection for lab-on-chip devices using external electrodes on the holder, *Analyst* **128** (8) (2003) 1019–1022.

[55] A. Berthold, F. Laugere, H. Schellevis, C.R. De Boer, M. Laros, R.M. Guijt, et al., Fabrication of a glass-implemented microcapillary electrophoresis device with integrated contactless conductivity detection, *Electrophoresis* **23** (20) (2002) 3511–3519.

[56] M.L. Reed, W.-K. Lye, Microsystems for drug and gene delivery, *Proc. IEEE* **92** (1) (2004) 56–75.

[57] H.J.G.E. Gardeniers, R. Luttge, E.J.W. Berenschot, M.J. de Boer, S.Y. Yeshurun, M. Hefetz, et al., Silicon micromachined hollow microneedles for transdermal liquid transport, *J. Microelectromech. Syst.* **12** (6) (2003) 855–862.

[58] E.V. Mukerjee, S.D. Collins, R.R. Isseroff, R.L. Smith, Microneedle array for transdermal biological fluid extraction and in situ analysis, *Sens. Actuators A Phys.* **114** (2–3) (2004) 267–275.

[59] R. Luttge, E.J.W. Berenschot, M.J. de Boer, D.M. Altpeter, E.X. Vrouwe, A. van den Berg, et al., Integrated lithographic molding for microneedle-based devices, *J. Microelectromech. Syst.* **16** (4) (2007) 872–884.

[60] B. Leboulanger, J.-M. Aubry, G. Bondolfi, R.H. Guy, M.B. Delgado-Charro, Lithium monitoring by reverse iontophoresis *in vivo*, *Clin. Chem.* **50** (11) (2004) 2091–2100.

[61] M.J. Tierney, J.A. Tamada, R.O. Potts, L. Jovanovic, S. Garg, Clinical evaluation of the glucowatch biographer: A continual, non-invasive glucose monitor for patients with diabetes, *Biosens. Bioelectron.* **16** (9–12) (2001) 621–629.

[62] E.X. Vrouwe, R. Luttge, A. van den Berg, Direct measurement of lithium in whole blood using microchip capillary electrophoresis with integrated conductivity detection, *Electrophoresis* **25** (10–11) (2004) 1660–1667.

[63] E.X. Vrouwe, R. Luttge, W. Olthuis, A. van den Berg, Microchip analysis of lithium in blood using moving boundary electrophoresis and zone electrophoresis, *Electrophoresis* **26** (15) (2005) 3032–3042.

[64] E.X. Vrouwe, R. Luttge, I. Vermes, A. Van Den Berg, Microchip capillary electrophoresis for point-of-care analysis of lithium, *Clin. Chem.* **53** (1) (2007) 117–123.

[65] Medimate Diagnostics. www.medimate.nl (accessed 12.04.11).

[66] B.A.P. Busher, U.R. Tjaden, J. van der Greef, Three-compartment electrodialysis device for on-line sample clean-up and enrichment prior to capillary electrophoresis, *Journal of Chromatography. A* **788** (1997) 167–172.

CHAPTER

Microfabrication for Novel Products in Drug Delivery: An Example

8

CHAPTER CONTENTS

In Chapter 1 the importance of technological innovation in the market place was discussed, and the previous chapter gave a successful example of technology transfer from university to spin-off. This chapter is dedicated to academic-driven product innovation. A new technology starts with results from various different scientific disciplines, together with an analysis of the existing state of the art in industry. One may suggest that there is already enough knowledge (business intelligence) in industry, and universities (operating on public money) are not the place that competitive products should

Microfabrication for Industrial Applications. DOI: 10.1016/B978-0-8155-1582-1.00008-3

be developed; they should stick to the provision of stimulating educational environments. Universities do, however, have a clear responsibility to take their place within the innovation chain. But what is that place exactly? Is it possible to provide the best academic training for new engineers without engineering sciences leading to products in the (competitive) industrial market place? I would immediately say that it must be both. With respect to modern learning, we believe that experience with real-world problems is almost as important to a student's education as instruction in the basics of his/her discipline.

Universities, therefore, have to create a motivating environment, in which the students actually get in touch with real-world problems, at least to a certain level. Normally, as an individual student and as a member of society, we easily recognize this need for gaining knowledge within universities. So, what is in this for industry? Since industry is simply the profit-making part of society, universities should bring their results to the attention of the industry, and similarly industry should bring their pre-competitive research questions (encountered in the *real world*) to the attention of academic scientists. Today, both actors operate together in a variety of extensive networks, in part stimulated by public money. Within these networks, an individual's effort can become linked to others and impact is created.

Well, can one expect that every individual's research effort makes a difference? This is unrealistic, but the sum of all the incremental results clearly makes a difference to society. Here, I would like to focus on a specific example, in which the results gained from researching a new technology are pre-competitive and open access, but the utilization thereof may well not be. Public resources must be used to benefit society as a whole, without discrimination and without private economical benefit to the individuals within the research teams. However, patents can still be obtained and the technologies described therein can be validated on industrial terms.

Technology transfer occupies the interface between knowledge creation and product development. Part of this process is the selection of which products to develop. Ideas which present immediate answers to specific problems in a market place have the highest chance of being developed; however, it is not always obvious which ideas will have value after a product development path which takes five, ten or more years towards completion prior to emerging on the market. For technology to emerge into the market place, visions are important.

This chapter focuses on the development of a new technology for drug delivery through the skin. In it, we will discuss state-of-the-art microneedle arrays, and a specific microfabricated device which can transport a drug or vaccine through the skin by means of nanoscale pores.

Before building a future scenario for the use of this type of device, I have to first thank the co-workers at University of Twente's MESA+ Institute for Nanotechnology, specifically Dr. S.N. Bystrova, M. Domanski and Dr. A.J.A. Winnubst, for their technological and scientific inputs in the development of a novel nanoporous microneedle array technology. The support of Prof. Dr. J.A. Bouwstra of the Leiden/Amsterdam Center for Drug Research (LACDR), Leiden University, is also acknowledged. She helped to define the real-world problem (enhancement of transdermal drug delivery) to us as engineering scientists and supported the valorization project through her expertise in the field of the characterization of skin-transport processes, and kindly suggested that fluorescent ovalbumin can be used as a model compound for the evaluation of the nanoporous microneedle array technology developed at University of Twente's MESA+ Institute for Nanotechnology spin-off MyLife Technologies. Further, I would like to acknowledge the financial support of the original research project, as well as the valorization project related to MyLife Technologies, from the Dutch Science and Technology Foundation STW, Applied Science Division of NWO, and the technology program of the Ministry of Economic Affairs, The Netherlands. Finally, I would like to mention that dedicating time and expert knowledge to the compilation of this chapter would not have been possible without the funding of my personal VENI grant, also by STW.

8.1 MICRONEEDLE RESEARCH AT UNIVERSITY OF TWENTE AND ITS SPIN-OFF

Chapter 7 discussed the LICETAS project. In this section, we will discuss novel products based on microneedle array technology (MNA) from the point of the commercial exploitation of high-tech research results by academically-driven product development. This route involves obtaining a patent and consists of two main elements: (1) desk research and (2) validating research for technology–product combinations. An effective academic-driven product development and innovation path always requires the collection of insight into the current needs of society. However, in science-driven technological development this path may take quite a long time in order to fully comprehend all its complexities. From idea, to proposal writing, to receiving the grant and hiring personnel takes approximately 1.5 years. Conducting the research and interpreting the results before being able to actually attract industrial attention may take another 2–3 years.

I have focused on the practical application of a specific microfabrication technology for microneedle arrays since 2006. This technology was developed by publically funded research at the MESA+ institute between 2003 to 2007 in collaboration with Leiden University. We investigated the enhancement of transdermal drug delivery processes within a pre-competitive research setting. In 2006, the project team at MESA+, headed by Prof. Dr. Ir. van den Berg (BIOS, The Lab-on-a-Chip group), made a patent application. To enhance the chance of utilizing the collected know-how, and to offer interested industrial partners (of which some were already part of the user committee consulting on the original research project) the opportunity to evaluate this technology further, the creation of a spin-off activity was initiated, and granted with a first phase valorization grant by STW of 25,000.00 Euros.

This money was required to transfer the know-how gained from the research results in the project into the first non-tangible asset of the spin-off: a patent application. Similarly to the LICETAS project, the first step in the project was necessary to secure a second stage pre-seed technology fund, which would then allow the technology to be developed to its full potential, thus reducing the economical risk involved in pre-clinical evaluation of the MNA technology. Based on the know-how built in the valorization team and the prospective patent application different business cases could be developed. This second, STW funded, valorization project (conducted from 2008 to 2010 within MESA+ at the research group Mesoscale Chemical Systems, headed by Prof. Dr. Gardeniers) increased the chances of a start-up within the highly competitive, but highly conservative medical field. This 200,000.00 Euro project allowed the team to validate the technology to identify a specific technology–product match, to create a first business niche for the novel patented nanoporous microneedle array product. We did not just look at the drug delivery market. To increase the opportunity for initiating a successful start-up, it is important to identify the unique selling points of the technology at hand. The publicly supported valorization project allowed us to bring the potential of the original research results to the attention of different parties with specific application needs. In November 2010, indeed,

The University of Twente's MESA+ Institute for Nanotechnology spin-off MyLife Technologies received an initial, first phase (50,000.00 Euros), life-science, pre-seed grant to evaluate the use of nanoporous MNA technology for making an ultra-minimally invasive, easy-to-use vaccination patch.

Through this process of pre-seeding, the organizational actors in the network could evaluate the business potential of the MNA technology, and support the efforts of the valorization team by the suitable instruments of technology transfer to the market. These efforts reduce the financial risks for an entrepreneur, allowing the public sector to get involved in the process of patent application roll-out to the national phase. At present, the MyLife Technologies team is negotiating the full terms and conditions of commercial exploitation of the patent with the university board and its funding agent. Bringing these negotiations to a successful close will enable the birth of a new company.

In the LICETAS case a company (Medimate BV) was started quickly after finalizing the original research project and know-how was transferred to this start-up directly. Why was this not possible in the MNA case?

In brief, the idea for LICETAS started from a project proposal inspired by the needs of a practicing psychiatrist, and funding was received in 2001. The PhD project was entirely dedicated to one specific appliation: quantitative lithium measurement from a droplet of finger stick blood. In 2005, a full proof-of-principle was presented to the user committee of the STW project, and at the same time it happened that a young entrepreneur was looking for a business case to start his own company. With the initial support of the original research team, and members of the user committee, it was decided to initially make the entrepreneur a part of the research team, through (1) an extension of 25,000.00 Euros of the original project and (2) the subsequent preparation of an STW valorization phase-one grant. Similarly to MyLife Technologies, it was because of this initial seed money that Medimate BV was able to take off. The entrepreneur was not part of the academic community; he had to provide a different instrument to become one of the actors in the network, by starting his privately held company (S. Staal BV) directly in the very first valorization phase. Eventually, after settling the knowledge transfer conditions with the university board and its funding agent STW, S. Staal BV and the other involved parties became owner of Medimate BV to continue the valorization path. Since this spin-off was created by the original research from STW funds, the University of Twente was also eligible to apply for funding in the second STW valorization round with its spin-off Medimate BV as the project conducting valorization team.

It is probably the case that a BV could not be started straight away in the MyLife Technologies case because different actors, at a different time, in a different location and in different circumstances of the actors' career paths were involved. Also, when the LICETAS research project had been finalized, no patent had been applied for, and the risk of shelving the project results was extremely high. Capillary electrophoresis on chip was not novel by itself, and detecting lithium by this technique in body fluids was not novel either, as it had been previously performed in serum. At the time, the accurate detection of lithium in a whole blood sample by capillary electrophoresis on chip might have had the potential to be patentable, but when the decision was made to transfer the know-how to the entrepreneur he already had a company in place, which increased the chances for utilization of the research results gained in the research project.

This scenario was not an option for MyLife Technologies, which brings me to another question: What can I do as an academic actor to manage the risk of shelving valuable research results versus the financially high risks involved in high-tech innovation? How can I conduct my academic research so as to generate output that is of direct value to a privately-held business that interacts with the academic network?

A comprehensive answer is probably not possible within the context of this book, hence I will restrict myself to the context of microfabrication for industrial applications. One future application of MNA technology may be the creation of a method for treating diabetes.

Today, insulin is given to a diabetes sufferer to control their glucose level. It is estimated that there will be 230 million diabetes cases worldwide in approximately 20 years from now, and today's methods are insufficient for providing a solution to this problem, so novel technologies will be required.

Many different researchers world-wide have been involved in the development of microneedle array technology for the ultra-minimally invasive delivery of insulin via the skin. Due to my involvement in publicly funded research on microneedle array technology for drug delivery and the patenting of a technology based on the research result we gained from the academic project, I was able to put my entrepreneurial head on and think towards a unique selling point for this technology in the identified application of insulin delivery. Is it a new business case? A significant amount of desk research was needed to answer this question. In the following sections I will describe its results and evaluate in depth a possible match for microneedle array technology with insulin delivery.

8.1.1 Desk Research: Microneedle Arrays, Microfabrication and Transdermal Delivery of Insulin

Microneedles are a next-generation drug delivery device, that targets the skin. As yet, few original research papers consider this type of insulin delivery, so a review of microneedle array microfabrication technology and its possible applications in transdermal delivery of insulin is presented. A small sub-set of publications has been identified that demonstrate the proof-of-principle for insulin delivery by microneedles. The results of these papers are compared. However, the specific criteria that a microneedle device must fulfil to be used in this way are unclear, and a variety of hypotheses can be derived from this desk research.

A variety of microneedle designs (tip shapes, length, diameter, materials, etc.) in three distinct design categories have been produced. These are: silicon micromachined, replicated and fine-mechanically manufactured arrays. Despite the fact that microneedles can be used to target the skin as a route for delivery in a nearly painless manner, none of the designs show an appropriate level of technological readiness for clinical applications. Many designs have been presented in the scientific literature, but so far, clinical evaluation is limited. To bring any of these minimally invasive devices to market, one has to develop a suitable model which allows clinical researchers to benchmark the emerging microneedle techniques with respect to the conventional hypodermic needle injection for insulin. The case study discussed here will highlight the essential research and development stages concerned with microfabrication of microneedle arrays.

8.1.2 Is There a Need for Microneedles?

An arsenal of specific microfabrication techniques has been developed for miniaturization technologies, of which photolithography has been the most successful. Inspired by the success of the microelectronics industry, a wealth of devices, not necessarily electronic, has been developed where small size is a prerequisite. Some contain both electronic and non-electronic components. These devices are called microsystems or micro(opto)electromechanical systems (MEMS, MOEMS). Since the 1980s, these architectures also have been explored for the manipulation of minute amounts of fluids, as has been discussed in preceding chapters. This activity has presented us with an entirely new field of research named microfluidics [1], which has made its mark in the life sciences by the introduction of so called Laboratory-on-a-Chip (LoaC) devices. These devices now present us with a multitude of applications, of which some are concerned directly with clinical applications [2]. One type of device is the microneedle array, either developed as a stand-alone device or integrated with sensors and actuators [3]. They may offer advantages over hypodermic needles or chemical penetration enhancers. For certain applications, whereby the target is actually in the skin, the direct insertion of active compounds into the viable epidermis will also increase efficacy of the therapy [4–8]. This is especially beneficial in the cases where the delivery of macromolecular compounds is desired, e.g., DNA vaccine [9]. The viable epidermis hosts antigen-presenting cells and these can be addressed by microneedle array therapy directly (see work by Birchall's group, Cardiff University, UK) [10–13]. Bouwstra's group at the Leiden/Amsterdam Center for Drug Research (LACDR), Leiden University, NL, presented two other examples in this area describing "Immune modulation by adjuvants combined with diphtheria toxoid administered topically in BALB/c mice after microneedle array pre-treatment" and "Microneedle arrays for the transcutaneous immunization of diphtheria and influenza" by Ding et al. [14, 15]. In some studies, microneedle arrays are used for skin pretreatment [16–19]. Other, more sophisticated miniaturized devices for drug delivery also exist [20–22].

Reviews of microneedle technology published to date [23–30] show they are virtually pain free compared to the hypodermic needle. Micromachined needles are intended to be limited in their penetration depth and should be painless in use because of their very sharp tips and defined length. Of course, passive patches and topical creams also avoid pain, however, they often work only for very potent, small molecule drugs, or in specific formulations with chemical penetration enhancers. Microneedle array devices, by contrast, create a multitude of highly-defined pathways through the stratum corneum, which allow the passage of molecules without the side effects of chemical enhancers.

Can this be applied to insulin delivery? When it comes to the micromanufacture of microneedles two distinct concepts exist: (1) the in-plane and (2) the out-of-plane approach, as depicted in Figure 8.1. During in-plane manufacture, the microneedle tips lie within the plane of the material. In the out-of-plane approach the microneedle tips evolve perpendicularly to the plane of the material's working surface, which offers two important advantages: (1) a huge number of parallel injection sites, and (2) seamless integration of micro/nanofunctionality from the needles into the back plate. The schematic top and the cross-sectional views shown in Figure 8.1 present

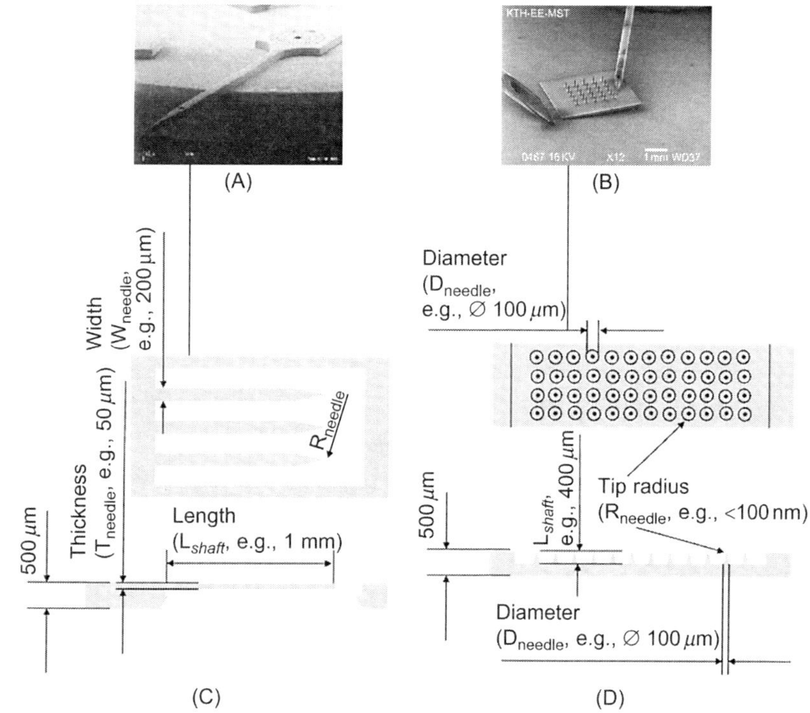

FIGURE 8.1

In-plane and out-of-plane micromachined microneedle arrays. (A) In-plane microneedles. (Reproduced with kind permission from reference [31]. Copyright ©2000 Springer Science+Business Media, LLC.) (B) Out-of-plane microneedles. (Reproduced with kind permission from reference [32]. Copyright ©2007 Springer Science+Business Media, LLC.) Schematic top and cross-sectional view of the in-plane (C) and the out-of-plane (D) microneedle designs, indicating the key design parameters for both concepts.

an overview of the key parameters which define needle geometry. While in-plane processing can deliver very long microneedles (L_{shaft}), the out-of-plane process is generally preferred because of its shallow, self-limiting insertion depth. However, the preferred technique depends on the individual application.

Combinations of these two approaches can be applied, for example Martanto et al. [33]. Microneedle fabrication uses the MEMS technologies, as outlined in Chapters 2 and 3, with the sequence of processing steps based on whether the needles are to be in-plane or out-of-plane, and can differ quite significantly per type of microneedle array (MNA).

The development of a clinically superior MNA delivery technology has been hampered by a limited understanding of the underlying transport mechanisms and the pharmacodynamics of the different tissue layers. The delivery of insulin, in particular, has been paid less attention than the more obvious applications of targeted delivery for vaccines and immuno-enhancing agents (adjuvants). Because diabetes is a severe disease with a rapidly growing patient population, it is important to investigate whether MNA devices possess distinctive benefits for its treatment.

In summary, this desk research provides a model case for entrepreneurial science and is organized in three sections. First a general literature screen of recent fabrication technology for microneedles is performed, including a discussion of the choice of materials and the technology requirements for commercialization. In the second section, a literature review of microneedle insulin delivery is presented. Thirdly, a sub-set of eight publications is derived, that describe successful experimental studies reaching proof-of-principle for insulin delivery by microneedles. In the latter section, specific attributes of these microneedle designs are discussed in more detail, which can eventually help to decipher the required criteria for the application of MNA to insulin delivery. Finally, the conclusion reflects on the state-of-the-art microneedle arrays for insulin delivery in clinical applications.

8.1.3 Microneedles by Microfabrication Technologies

Research Status

Pioneering research in the field of high-precision out-of-plane microneedle array fabrication and their medical applications were first studied by research groups at the Berkeley Sensor and Actuator Center, University of California (Berkeley, CA, USA), the Royal Institute of Technology (Stockholm, Sweden), MESA+ Institute for Nanotechnology, University of Twente (Enschede, The Netherlands) and the Georgia Institute of Technology (Atlanta, GA, USA) [3, 34–41]. A large number of publications have appeared over the last 15 years concerning the fabrication and testing of microneedles in a diversity of applications, including brain–computer interfaces, microneedle probing and stimulating for various brain regions, scalp electroencephalography, drug delivery, diagnostics and applications where spiked microstructures are used as nozzles and pipettes.

MNAs containing microneedles of diverse shape, length, diameter and density have been demonstrated to be capable of bypassing the skin barrier.

Both in-plane and out-of-plane fabrication are well suited to the creation of microneedle arrays, but each requires a different approach if it is to be integrated into a skin patch. Only the out-of-plane method will directly produce a two-dimensional array seamlessly connected to a back plate, which is preferable for integration into higher functionality [3]. MNAs are also fine-mechanically manufactured by assembly from conventionally produced hypodermic needles, and successfully demonstrated in certain types of transdermal delivery studies [42]. Nonetheless, lithographic exposure techniques utilizing microsystems technology (MEMS), (i.e., reproducing a structural element multiple times from a mask or mold), are now established in drug delivery research.

Lithographic techniques result in arrays with a tip radius of <100 nm, microneedle outer diameters below 300 μm, very well-defined length in the range of a few to hundred micrometers, and varying needle densities. Batch processing allows some cost reduction. Hollow MNAs require a flow-through delivery system and have been tested by gluing the back plate of the MNA to a syringe. These devices are not yet applicable to self-administered transdermal insulin delivery, but they demonstrate a clear step forward. Figure 8.2 summarizes complementary examples of out-of-plane MNAs, and demonstrates early prototypes of microneedles coupled to a syringe for intradermal drug delivery [33, 43–46]. Recent publications have discussed microfabricated microneedle technology, and evaluated their performance against other insulin delivery techniques [24, 47–49].

Fabrication and Design Concepts

Some processes form the MNA device directly, while others first produce a master. Silicon micromachining and stereomicrolithography are two examples of direct techniques forming three-dimensional work pieces [3, 37, 51]. SU-8 photolithography on pre-patterned silicon substrates combining lithography with micromolding, titanium micromachining, and the use of two photon induced polymerization of organic–inorganic hybrid biomaterials [52–54] have also been attempted experimentally.

Some remarkably simple methods have been recorded for micromolding microneedles. For example, Lee et al. introduced drawing lithography for the three-dimensional fabrication of an ultrahigh-aspect-ratio microneedle, and Zhu et al. recently demonstrated a fabrication process that uses chemically etched silica needles as a template for a replication process [55, 56]. Unfortunately, Zhu et al. [56] omit a detailed description of the fabrication

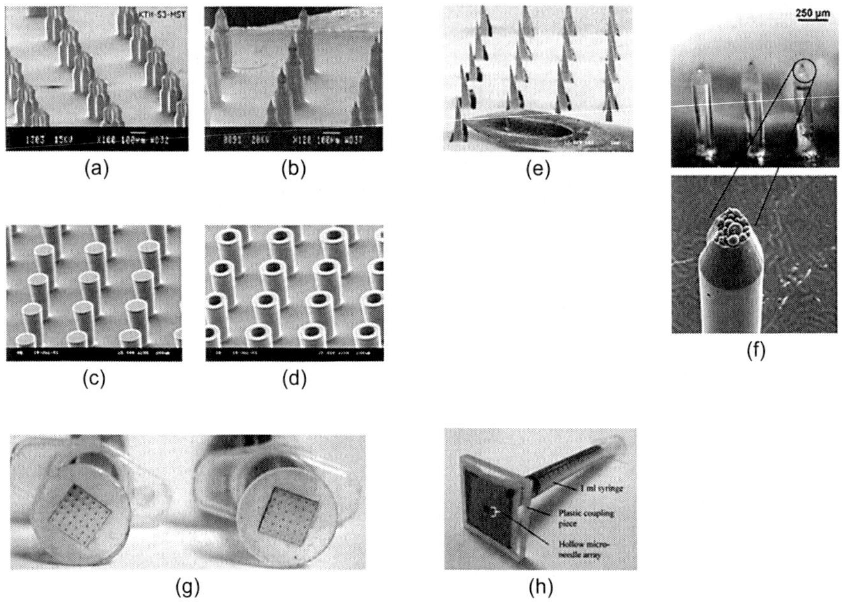

FIGURE 8.2

Different types of out-of-plane microneedle array technologies used for transdermal applications. (a, b) Sharp-tip silicon microneedle arrays. (Reproduced with kind permission from reference [43]. Copyright ©2008 Springer Science+Business Media, LLC.) (c, d) 80 and 100 μm diameter solid and hollow microneedle arrays without sharp tips. (Reproduced with kind permission from reference [44]. Copyright ©2005 Springer Science+Business Media, LLC.) (e) Stainless steel laser cut microneedle arrays. (Reproduced with kind permission from reference [33]. Copyright ©2004 Springer Science+Business Media, LLC.) (f) Two different types of dissolving polymer microneedle arrays by micromolding. (Reproduced with kind permission from reference [50]. Copyright ©2006 Springer Science+Business Media, LLC.) (g) Device prototypes used for clinical pain testing. (Reproduced with kind permission from reference [46]. Copyright ©1996 Springer Science+Business Media, LLC.) (h) Device prototype with 10 by 10 hollow silicon microneedles as shown in (d). (Reproduced with kind permission from reference [44]. Copyright ©2005 Springer Science+Business Media, LLC.)

of the silica template. Although their paper does not describe the fabrication method of the template the authors clearly identify micromolding as a low-cost manufacturing scheme, which is suitable for further drug delivery research.

Many different processes have been tried to make extremely sharp microneedles. Unfortunately, there is no consensus on how one should define sharpness in this context. Many different research groups have elaborated on various fabrication processes to generate particular needle geometries, with the intention of reducing the tip radius, which will obviously lower the insertion energy. The density, the total number of needles within a defined array area, and the mechanism of insertion (manual or impact) are likely to be the major factors influencing insertion performance. Knife-like features, such as that introduced by Gardeniers et al. [38], should be assessed against other designs with respect to pain sensation, skin reaction, intrinsic skin variations from volunteer to volunteer, place of incision and relevant parameters of wound healing in relation to the size and geometry of the incision. Development of a benchmark for these parameters is not trivial because the skin is an anisotropic biomaterial, which changes its mechanical characteristics under stress. Moreover it is unclear how the tip radius of a microneedle (i.e., sharpness) can be translated into a meaningful parameter for such an incision experiment.

Some studies have explored needle radius together with other geometrical and application parameters through mechanical modeling [57, 58]. However, the ease of insertion of the microneedles or the successful introduction of drug does not just depend on the individual needle tip, or the configuration of the array, but also on the apparatus that is used to administer the MNA device to the skin and the drug formulation [59]. Furthermore, a trade-off between sharpness and microneedle stability must be sought for a clinical application. The insertion of microneedle arrays in human subjects was reported in a study by Gill and Prausnitz [60].

Materials

Microneedles are made either directly from active agents embedded in a dissolving material, or the microneedle platform acts as a carrier, which releases active compounds from a coating (solid microneedles), through a fluid conduit from a reservoir (hollow microneedles) or from the material's intrinsic pores (porous microneedles). The range of materials used in research or commercial activities includes glass or silica, glass-like plastics and ceramic composites, silicon, silicon nitride, stainless steel, nickel-plated needle arrays and titanium, amongst various types of polymers, carbohydrates and micro-nanoparticle hybrid materials. Each of these materials may require their own fabrication techniques, for example, stainless steel or titanium microneedle arrays are cut by a laser, which limits their physical dimensions to the resolution of the laser cutter. This limited resolution will also affect the radius of the needle tips, and the overall geometrical dimensions of the needles.

Silicon, which forms a native oxide layer, is much used in microsystems technology and has been accepted for use in transdermal pre-clinical research, although no commercial application yet exists. Ji et al. have done some preliminary work in the area of porous microneedles. They used selective electrochemical etching, with a photoresist as a protective mask to form the porous silicon tip, thus forming one of the first silicon out-of-plane microneedle arrays that contain tips of two types of materials with very distinct properties [61]. Other publications also discuss the range of materials used in out-of-plane micro- and nanomanufacture of MNAs. For example, Park et al. investigated a polymer double-layer molding process, as shown in Figure 8.2f, which was generated from a mold structure. The mold was fabricated by means of a combination of SU-8 photolithography, reactive ion etching and polydimethylsiloxane soft lithography [45, 50]. They produced tapered tips, using poly(lactic-co-glycolic acid) as a material for the formation of dissolving microneedles and encapsulated calcein in the tips. The particle structure remains intact inside the molded microneedle.

Similarly, Moon et al. fabricated a mold by soft-lithography, from a master made by inclined deep x-ray exposure, then the soft-mold is used for the replication of MNAs into biodegradable poly-L-lactide acid [62]. Carbohydrates have also been used for molding MNAs, and Donnelly et al. discussed the processing difficulties and instability of the material in MNA production [63]. Kolli and Banga, and Li et al. have found maltose to be more easily adapted to molding microneedles [64, 65]. Both Kolli and Li are members of Banga's group affiliated with the Mercer University (Atlanta, GA, USA), which holds an impressive record on microneedle array technology.

Ceramic Nanoporous Microneedles

Within microneedle research, some groups are attempting to make ever sharper and smaller point-like microfeatures, while others are looking further on in the development chain, towards low-cost production and applications. One particularly exciting concept is the possibility of making arrays of microneedles in one platform, so combining a multitude of microneedles in parallel without the need for additional assembly steps. Such arrays could be used to make hundreds of miniscule incisions into the skin in one step, without causing pain. These incisions could then be used to transport molecules across the skin barrier. Figure 8.3 shows a ceramic-based nanoporous MNA platform technology developed at University of Twente's MESA+ Institute for Nanotechnology (patent pending WO2009/113856). Here, the MNA is replicated from a soft-mold that is generated by polydimethylsiloxane

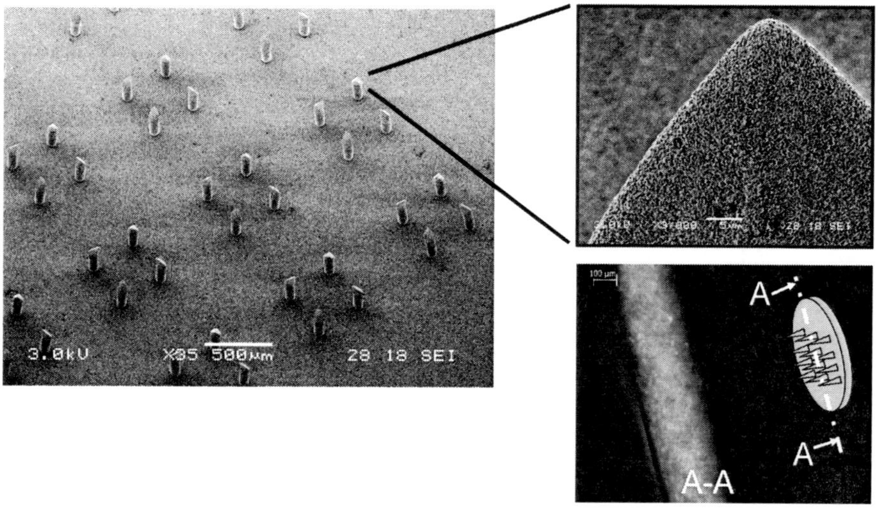

FIGURE 8.3

Scanning electron micrograph of a nanoporous microneedle array developed at MESA+ Institute for Nanotechnology, University of Twente. Inset (top right) shows a scanning electron micrograph revealing the porous structure of an individual needle tip. Inset (bottom right) depicts an optical micrograph showing a schematic of the array and its cross-section, demonstrating that the resulting nanopores can be filled with fluorescent ovalbumin (45 kDa), e.g., by incubating the nanoporous ceramic material in the protein solution over night. (Image courtesy: S.N. Bystrova, University of Twente, 2009.)

soft-lithography from a master – this technique was discussed in Chapter 3, Section 3.3.4. A specific master was designed and manufactured by a combination of SU-8 photolithography and silicon micromachining that also integrates knife-like features in the tip with similar dimensions as introduced in previous work by Gardeniers et al. [38].

A nanoporous microneedle has a porosity of approximately 30%, and an average pore size of 100 nm (see inset Figure 8.3, top right), which is large enough to hold model molecules such as fluorescent ovalbumin (inset Figure 8.3, bottom right). Further investigation is necessary to determine the extent to which such non-dissolving, nanoporous platforms can be of use in the delivery of macromolecules such as insulin.

In general, transdermal MNA devices are classified by the Food and Drug Administration as an invasive device, since the skin barrier is bypassed and also probably ruptured at deeper layers than the stratum granulosum. In most cases applying such devices to the skin surface will allow substances other than the drug to enter the body. Any skin rupture event will interfere with the body's defence mechanism, generating additional adverse reactions in the skin, or even unfavorable systemic effects through infection or poisoning, hence this classification is necessary. All components of the microneedle administration device that come into direct contact with tissue, or risk contact during operation of the device must, therefore, obey stringent rules to gain medical approval.

At this stage of development, the exact material properties of a MNA are not fully understood, and have to be assessed as they stand and also when they are packaged within a drug-applying device. Despite their general potential being recognized in the drug delivery community the benefits of a specific material for insulin delivery still have to be addressed for all the different novel platform technologies described above.

Figure 8.4 gives an overview of the nanoneeedle fabrication process developed at University of Twente's MESA+ Institute for Nanotechnology. This process starts with a hard mold, manufactured by photolithography and silicon micromachining, to combine a pre-patterned silicon substrate and a thick-film photoresist (SU-8) into one hard-mold structure for the fabrication of a soft mold for subsequent multiple production of ceramic green bodies. The individual MNAs are cut from the green body and sintered, forming the semi-finished product ready for integration into a delivery patch.

Figure 8.5 depicts the integration of the ceramic MNA into a patch-type system. With a resulting materials porosity of approximatly 30%, a single microneedle within the array has a liquid loading capacity in the order of 1–2 nl, and the backing plate can contain up to 5 μl. These minute amounts may suffice for vaccination purposes. Currently, the application of University of Twente's MNA technology is being investigated in collaboration with Prof. Dr. Scheper's and Dr. de Gruijl's groups at the VUmc Free University Medical Center, Amsterdam, The Netherlands, for the purpose of vaccine delivery. This type of nanoporous MNA is probably not appropriate for insulin delivery, but one needs further research to be conclusive.

Summarizing the Technology Requirements from a Commercialization Point of View

Microfabricated devices compete with the delivery of insulin by standard, ultra-fine needle systems, either integrated in a patient-friendly, pen-like delivery system, or as a classical hypodermic needle mounted on a syringe. This type of needle costs about 0.05–0.20 US Dollars per disposable needle.

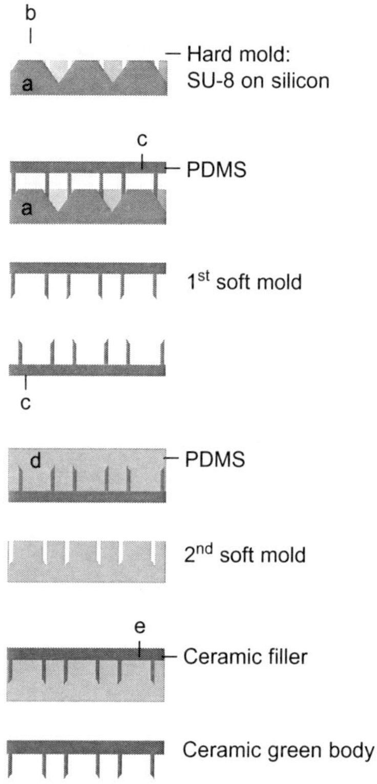

FIGURE 8.4

Process sequence of micromolding for ceramic microneedle arrays including the fabrication of a soft mold for production. Patent pending. (Explanation of a–e can be found elsewhere, [patent pending WO2009/113856])

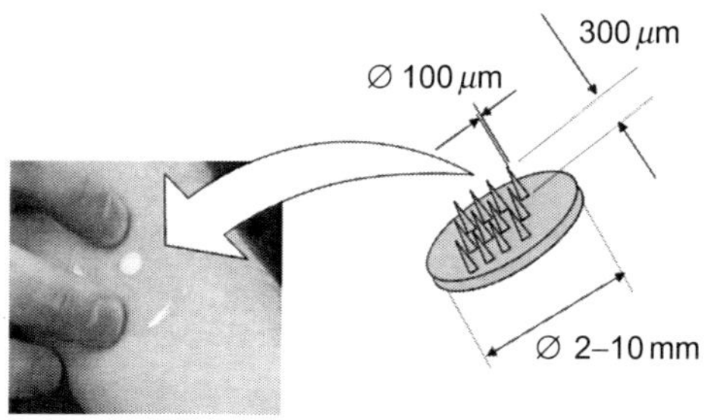

FIGURE 8.5

Envisaged ceramic microneedle array integrated skin patch.

MNA replication technology can probably compete with such a low pricing regime, but the pharmacodynamics of insulin uptake from the epidermal layer needs much more research, and is not yet clinically identified as a beneficial route for insulin delivery.

Small-molecule-releasing skin patches are already far better understood, and have been marketed at a large volume in industrial pharmacy. Fitting microneedles into one of the existing skin patch platforms may accelerate its integration into a technological solution. An example of successful, interdisciplinary, transdermal, drug delivery co-development is evident in current products in the iontophoretic patch market. These products combine microelectronics integrated current-controlled power supplies with skin patch technology into a patient-controlled iontophoretic transdermal system, providing on-demand systemic delivery of substances such as fentanyl (Ionsys™) and lidocaine (LidoSite®) [66]. Clinical trials are under way for both these systems. These electrically-controlled systems are quite complex, so it is surprising that passive microneedle arrays in a skin patch lag behind them in the market.

Another very promising example of MNA technology in a transdermal application is being commercialized by Zosano Pharma Inc. Zosano Pharma's Macroflux® transdermal microprojection delivery system combines the convenience of a needle-free administration process with room temperature stability for various therapeutic peptides, proteins, small molecules and vaccines. The Macroflux® microneedle array component has been described in the scientific literature for more than 10 years.

Upgrading MNA technology from a device that just creates microscale aqueous conduits into the skin barrier into a fully packaged and safe drug delivery system comparable to Macroflux® will clearly require more than just engineering ingenuity. A device platform must be designed and manufactured, that will not only allow testing of a small number of demonstrators for transport studies, but must also present a better solution for diabetes management, justifying patient trials supported by the shared risk of private–public partnerships. Once such a microneedle technology platform is confirmed by evidence-based clinical translational research, it can be re-engineered into a more flexible system that will accommodate and address individual patient's needs through the appropriate industrial channels for market introduction. The state of the art in microneedle array technology as it exists for insulin delivery in the scientific literature is discussed in the next two sections.

8.1.4 Are Microneedles Ready for Insulin Delivery?

In the last two decades a number of papers discussing the fabrication of microneedle arrays, either starting from the in-plane or the out-of-plane manufacturing concept, have been published. An evaluation of their application to insulin delivery is described in this section. One of the earliest papers addressing insulin delivery by means of MNA is the publication "Solid and hollow microneedles for transdermal protein delivery" by McAllister et al. [40]. Solid silicon microneedles were inserted across the stratum corneum and painlessly disrupted the skin's barrier function, as demonstrated by an increase in permeability to calcein by four orders of magnitude. Additionally, an *in vitro* experiment was carried out that delivered insulin across the human epidermis at approximately four orders of magnitude larger levels than passive permeation.

These experiments utilized solid microneedles, and were carried out as a microneedle pretreatment intended to generate aqueous pathways into the epidermis. The publication also addressed fabrication concepts for hollow silicon and metal (nickel) microneedles, producing out-of-plane microneedles that permit fluids (water) to flow through their bores for the first time. Although this work was highly experimental, it provided a clear milestone in the field of microneedle array fabrication technology. The two pioneering researchers in this field were A. Pisano and M. Prausnitz. Although this review focuses on out-of-plane microneedles, Pisano and collaborators have also published an in-plane polysilicon micromolding process that was developed to fabricate hollow tubes by a combination of various micromachining techniques (see Figure 8.1(A) in Section 8.1.2). The researchers claimed that these types of devices could be useful for microfluidic applications, specifically including the continuous delivery of insulin to a diabetic patient [31]. In 2000, Zahn et al. (Pisano's group) also considered the use of their microneedles for sample collection in biological analysis [31]. Their findings particularly mention microneedle strength and achievable bending moments, and optimization thereof. They further studied the flow capacity of their microneedles given a pressure at the inlet up to 138 kPa. Despite their application claim, no results on insulin delivery were described at this stage of their research. In the same year an overview of microneedle technology was published by the Prausnitz's group which extended the list of possible microneedle applications to gene and macromolecular drug delivery, and that may include insulin [67].

Trebotich et al. then presented a thorough study of the geometries and flow characteristics of microneedles [68, 69]. Although the analysis of in-plane design differs from the required analysis for out-of-plane designs, it is worth mentioning that these mechanical parameter studies should be translated to experiments evaluating skin insertion performance. Zahn et al. have also published "Components of an integrated microfluidic devices for continuous glucose monitoring with responsive insulin delivery" in Diabetes Technology and Therapeutics [70], in which they suspect that their microneedle drug delivery technology could be applied to the intradermal injection of insulin. They hypothesized that intradermally injected insulin is quickly absorbed by the capillary bed into the bloodstream – an important postulation that still requires further clarification. They also postulate that the microfluidic administration of insulin could be performed by an integrated active device and may therefore allow complex drug delivery profiles compared to the classical bolus injection of insulin.

The methods under investigation to pass the skin's barrier have been reviewed, as have fabrication methods and transport studies of microfabricated needles [41, 71]. Amongst other large molecular weight compounds and nanoparticles, this study included the first proof-of-concept that showed the flow of microliter quantities of insulin into skin utilizing an individual hollow glass microneedle *in vivo*, and reducing the blood glucose level in diabetic rats.

The fabrication review was a follow-up of the preliminary work carried out on hollow microneedles by McAllister et al. in 1999 [40]. Based on silicon micromachining, a novel out-of-plane microneedle fabrication technology was demonstrated by Gardeniers et al. in 2002 [38]. In their follow-up publication in 2003, they demonstrated *in vivo* insulin delivery into rats for the first time utilizing these hollow silicon microneedles coupled to a syringe [3]. Prausnitz and his various co-workers have also continued their work on insulin delivery at different levels of integration. For example, in 2004 Martanto et al. described laser-cut, stainless steel microneedles for transdermal delivery of insulin *in vivo* by a solid microneedle pre-treatment [33]. Prausnitz then showed that microneedles can dramatically increase transdermal delivery rates, especially for macromolecules [24]. These findings confirmed the earlier postulation of Zahn et al., although the mechanism of insulin uptake still remains unclear [70]. Together with an overview of proven transport studies, including the example of insulin facilitated by microneedle pre-treatment, Praunsnitz' findings incorporate the concern of skin insertion force (i.e., margin of safety). Prausnitz also extended his view on advanced delivery mechanisms that could be achieved by combining microneedles with iontophoresis. To the best of my knowledge, this concept was first demonstrated for insulin delivery by Chen et al. in 2009 [17].

Inspired by the side-open microneedle design introduced earlier by Griss and Stemme [37], Zhang and Jullien described "Microneedle arrays for drug delivery and fluid extraction" including a side-open microneedle out-of-plane design with very sharp points, optimized through the use of a bi-mask technique [72]. They claimed that their needles are useful for insulin delivery into the human body, amongst other applications. Unfortunately, they did not include any results of a study on humans. In 2009, they used the same microneedle array design to demonstrate ink delivery into chicken skin [73]. Similarly to Gardeniers et al., Teo et al. coupled their microneedle arrays to a syringe [3, 44] and performed experiments with straight-walled microneedle arrays made from silicon. These results are of a controversial nature, since they used blunt microneedles despite all other systems using pointed ones. They demonstrated an increase in transport rates through dermatomed skin, while their findings *in vivo* restrict the potential benefits of this MNA design. Teo et al. state the following in their conclusion:

> "In vitro *tests using isolated skin demonstrated that the application of the microneedles resulted in a 10–20 fold increase in transdermal transport depending on the diameter of the microneedles.*"

Furthermore, Teo et al. concluded:

> "In vivo *test in diabetic animals, however, demonstrated that arrays of these microneedles were not effective in the delivery of a systemic drug, such as insulin.*"

However, in the same year an additional study by Davis et al. (Prausnitz's group) presented further work on insulin delivery enabled by the microneedle array technology introduced by McAllister et al. [74, 41]. These contradictory findings may be explained solely by the difference in the geometry of the needles, their array density and the experimental differences between the set-ups of the *in vivo* trials.

In 2006, other work on MNA technology for transdermal insulin delivery was published. For example, Ma et al. introduced "A PZT insulin pump integrated with a silicon microneedle array for transdermal drug delivery" [75], but this research does not demonstrate any insulin delivery either *in vitro* or *in vivo*. Lv et al. published "Modeling of transdermal drug delivery with a microneedle array", in which their theoretical studies were directed at insulin delivery [58]. Computer modeling of the fluid injection performance from a MNA is an important design aspect in the development of such applications. For example, transport into the skin depends on factors such as injection velocity and blood circulation. These initial modeling efforts are extremely valuable in optimizing this class of devices for clinical practice, and they must be continued as an integral part of product development.

Roxhed et al. have presented a higher level integration concept for drug delivery by introducing "Compact, seamless integration of active dosing and actuation with microneedles for transdermal drug delivery" [76], giving the example of insulin, which has been eventually proven in a publication by Nordquist et al. entitled "Novel microneedle patches for active insulin delivery are efficient in maintaining glycaemic control: an initial comparison with subcutaneous administration" [32]. This paper provides the first proof-of-principle of insulin delivery by an integrated microsystem that uses a hollow silicon microneedle array in the form of a patch.

At the same time, Ito et al. introduced their microneedle concept for the percutaneous absorption of insulin, in the *European Journal of Pharmaceutical Sciences* [77]. In this study, a total of five microneedles delivered insulin into the skin. The microneedles were fabricated forming a thread with polypropylene tips. This technique is not comparable to the high-precision microfabricated array described previously, but their work eventually led to the development of micromolded, self-dissolving microneedle arrays for insulin delivery, introduced in 2009 [78].

Just one year before, Roxhed et al. presented controlled administration of insulin in their publication on "Painless drug delivery through microneedle-based transdermal patches featuring active infusion" in rats [79], which complemented their earlier studies on "Membrane-sealed hollow microneedles and related administration schemes for transdermal drug delivery" [43], which covers a specific design element of a dissolving or bursting membrane at the tip, gaining a higher integration level and thus potentially more control over the various parameters involved in the delivery mechanism. A relatively new review by Prausnitz and Langer published in 2008 two additional research papers, and discuss how insulin delivery depends on MNA geometry. They were followed in 2009 by "Optimizing microneedle arrays for transdermal drug delivery: extension to non-square distribution of microneedles" [57, 80, 81].

These three papers are entirely dedicated to modeling and their findings must be evaluated further in clinical translational experiments. Finally, a relatively recent study by Gupta et al. shows data that compare human *in vivo* studies by means of microneedles with bolus injections [49]. Although this very recent study by the Praunsnitz's group uses a single pulled glass needle for the insulin delivery experiments, as opposed to their earlier out-of-plane hollow microneedle arrays, it is still important in evaluating the key parameters of microneedle insulin delivery. In summary, only a few significant out-of-plane microneedle technologies have been presented throughout the last decade. Figure 8.6 depicts the three design categories. Category (I) illustrates direct silicon micromachining from top to bottom: blunt needles by

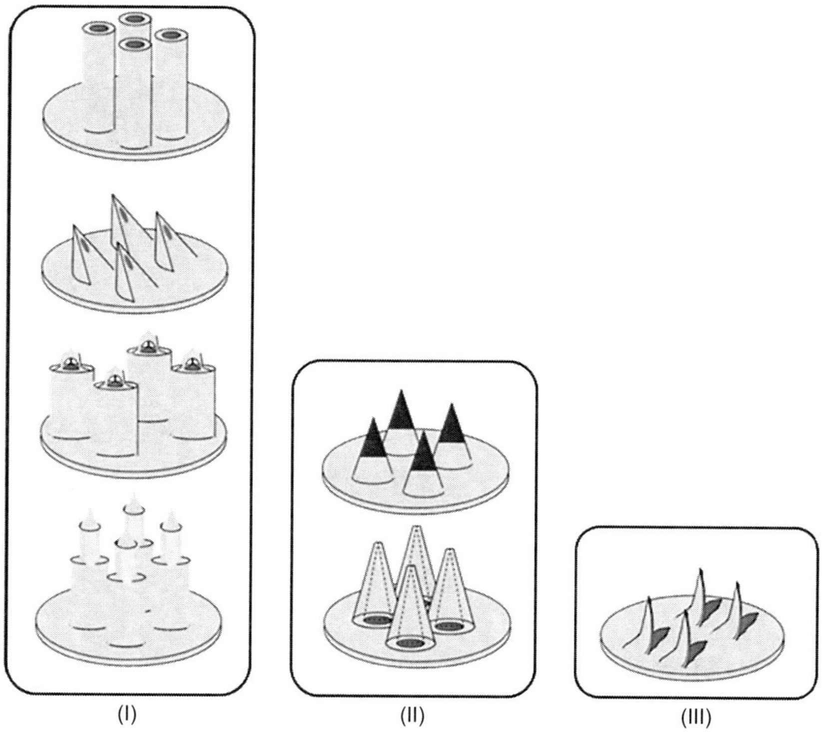

FIGURE 8.6

Design categories: (I) direct, (II) replication, (III) fine-mechanical approach. The drawings reflect on the alternative designs. The generated shapes and dimensions are strongly linked to a selected fabrication process. All design varieties have proven to rupture the skin. Every specific shape, i.e., fabrication process, has dimensional restrictions.

Teo et al., sophisticated knife-like needles by Gardeniers et al., very pointy needles with side open release channels introduced by Griss et al., optimized by Roxhed et al., and finally utilized patches in insulin delivery experiments by Nordquist et al. [44, 3, 37, 76, 32].

Category (II) introduces replication processes containing two extreme examples of this category: a dissolving polymer microneedle introduced by Ito et al. [78] and the metal microneedles described by Davis et al. [74]. Category (III) illustrates one of the examples of a fine-mechanical approach. This needle is manufactured by laser cutting and bending of a metal sheet. This type of microneedle has been used as a pre-treatment device, by means of which insulin delivery in a diabetic rat model has been confirmed [33]. All these designs are inherently different, however, they all give a proof-of-principle for microneedle insulin delivery. These design differences can only be further evaluated within an application-oriented design process.

As an example given on insulin, innovation in medical interventions is focused on the development of more effective therapies, having less side effects and lower costs. Injection by a needle-and-syringe system for the administration of drugs or vaccines is an example of such a medical intervention. Although this method is well established and very effective in medical practice (inserting a molecular substance in to the body that normally would not pass the skin passively) it does have drawbacks. The most obvious is that a needle generally causes pain, when it contacts the sensitive nerves inside of the skin sending signals of pain to our brains. Therefore, a variety of other mechanisms of transdermal delivery and micromachined microneedle systems have been invented, including microneedle patches as described above for the delivery of insulin. Figure 8.7 gives another example of such a transdermal application, whereas a commerically available microneedle system acts as a skin portal, which is developed by Zosano Pharma Inc. (previously ALZA Corporation).

8.1.5 Design Aspects for Microneedle Insulin Delivery

At present, different types of insulin formulations are available with different pharmacological specifications. They are needed because patients require both a basal level of insulin and rapid-acting insulin to deal with food intake and exercise. Clearly, to be useful to patients, microneedle delivery must be able to supply both long- and short-acting insulin supply. Eventually, clinical scenarios could include the use of different microneedle configurations and the use of different insulin formulations.

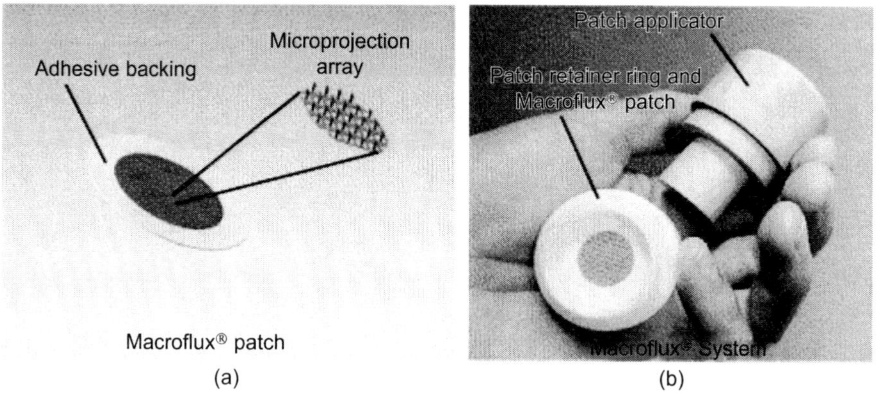

FIGURE 8.7

Macroflux® delivery system. The patch comprising the coated microneedle array affixed to an adhesive backing is illustrated in figure (a). The patch loaded on the disposable retainer ring and the reusable applicator are illustrated in figure (b) with its applicator device. (Reproduced from reference [82].)

However, as yet, only eight publications were found that confirm the proof-of-principle of using MNAs in insulin delivery in the diabetic rat model, while one additional publication [33] discusses microneedle insulin delivery in human subjects of Diabetes Type 1 using a single glass microneedle [3, 17, 32, 43, 44, 74, 78, 79].

Fabrication Attributes and Their Clinical Relevance

Shaping of the MNAs directly, by silicon micromachining techniques (category I, Figure 8.6), is considered to be an expensive technical solution due to the relatively high cost of high-precision silicon wafers as a base material. These techniques allow a high level of integration, and are advantageous when a certain amount of electronic integration is needed for the device. These processes will also be capable of producing very sharp tips, with radius less than 100 nm. This reduces the insertion force needed for such devices, thus enabling ease of use for manual administration. A first comprehensive trial, benchmarking important delivery parameters in the diabetic rat model, was carried out by Nordquist et al. [32]. Figure 8.8 shows two of their findings: (1) the decreasing plasma glucose level, showing the effect of the route of administration by microneedles (Figure 8.8a), and (2) the effect of the rate of intradermal administration on blood glucose (Figure 8.8b).

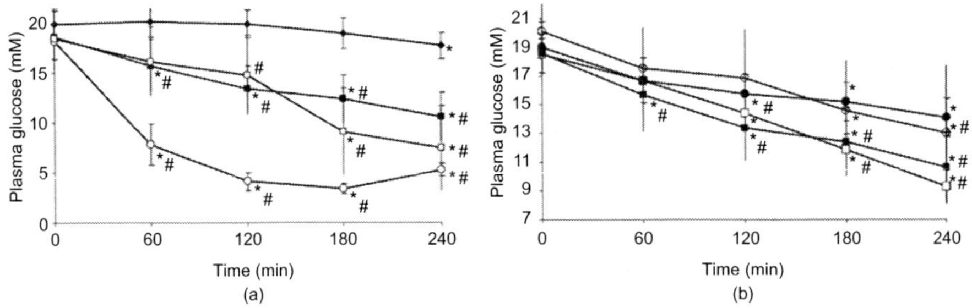

FIGURE 8.8

Plasma glucose measured in the diabetic rat model. Asterisk(*) denotes a difference compared to the first measurement in that same group and number sign(#) denotes differences when comparing to vehicle-treated animals within the same time period. For all measurements, $P < 0.05$ was considered statistically significant. (a) The effect of route of administration of insulin lispro on blood glucose. Filled square denotes microneedle-aided intradermal infusion (intradermal low-rate delivery rate, 2 μl/h, 100 IU/ml, $n=9$), open square denotes subcutaneous infusion (2 μl/h, 100 IU/ml, $n=9$), filled diamond denotes time control, and open circle denotes intravenous infusion (2 μl/h, 70 IU/ml,$n=7$). (b) The effect of rate of intradermal administration of insulin lispro on blood glucose. Filled square denotes microneedle-aided low-rate intradermal infusion (2 μl/h, 100 IU/ml, $n=9$), open square denotes microneedle-aided high-rate intradermal infusion (4 μl/h, 100 IU/ml, $n=9$), filled circle denotes microneedle-aided intradermal passive infusion (0 μl/h, 100 IU/ml, $n=9$), and open circle denotes rats whose skin was penetrated with microneedles, and then passively infused (2 μl/h, 100 IU/ml, $n=9$). (Figures and caption adapted from Nordquist et al. by permission of the author and the publisher. Reproduced with permission from reference [32]. Copyright ©2007 Springer Science+Business Media, LLC.)

The second category (II), Figure 8.6, covers replication processes from a master. In general, these are cheap production schemes at a relatively low level of integration. Polymer micromolding processes incorporating insulin into a matrix material were introduced by Ito et al. This group used a relatively expensive, lithographically-defined master to yield very well-defined features in a biodegradable matrix [78]. As an alternative, batch-replication by electrodeposition from a relatively cheap sacrificial master, made by laser machining is less accurate but generates a very specific design with a through hole [74]. These two process types may each have benefits and opportunities for cost reduction. From the affordability point of view, high-volume replication from a sustainable master structure is the most feasible route.

Clinically, however, these two replication concepts are very different. The dissolving polymer microneedle can only allow a fixed amount of insulin release per time interval. On the other hand, the hollow metal microneedles can be coupled to an insulin reservoir [74, 78], which can in turn be linked to a closed-loop feedback sensor, which triggers release of insulin on demand. This is similar to continuously controlled release from an implanted or percutaneous drug delivery pump, and thus should bring high patient compliance. Such concepts anticipate high efficacy and reduced side effects during therapy. Other hollow micromachined microneedles, e.g., as given in category (I), also offer this kind of advantage.

The devices in category (III), Figure 8.6, are made by fine-mechanical machining and are perfectly suitable for a relatively small number of devices that bring a technical solution to a drug delivery niche market, but they remain rather expensive for high-volume manufacturing, so their widespread application is limited. A variety of studies are described that use fine-mechanically manufactured out-of-plane microneedles. Martanto et al., for example, experimented with different insertion configurations [33]. The microneedles were inserted by a specific impact device and then removed after 10 s, 10 min, and 4 h, respectively. Leaving needles in the skin for longer led to smaller reductions in blood glucose levels, and multiple insertions of the device created more pathways into the skin. In these studies the area of skin that was treated with microneedles was covered with an insulin-containing reservoir. Although some variations in transport were observed during these different application procedures, the general conclusion was that the microneedles created aqueous pathways. Utilizing this fairly simple pre-treatment mechanism, an overall pharmacodynamic response to insulin delivery was demonstrated, in which blood glucose levels decreased in a similar manner to subcutaneous hypodermic injection.

Table 8.1 compares the different microneedle configurations that were used in the rat model, with reference to the parameters illustrated in Figure 8.1. So far, there is no clinical evidence that the more precise shapes lead to a higher delivery efficacy. The device design depicted in category (III) of Figure 8.6 allows very simple and straightforward integration with a drug-filled adhesive patch by lamination [83]. The patch releases its cargo through openings in the back plate. The release profile will depend on the interplay of the type of incisions made by the spikes with the opening in the back plate, and varies quite significantly amongst different skin types or age groups of patients. It is difficult to say how robust the different design categories actually are for insulin delivery in the real world, and if they could produce the best type of incision. Further clinical research is required.

Table 8.1 Microneedle Arrays for Insulin Delivery in the Diabetic Rat Model

Ref.	Material, R_{needle} (nm)	D_{needle} (μm)	Width (μm)	Thickness (μm)	L_{shaft} (μm)	Needles/ array	Animal	Procedure	Type of insulin	Response
3	Silicon	Di, ellip. 70	250 (at base)	–	150 and 350	81	Sprague-Dawley	Hollow microneedles	Humulin-R, 1 IU/h	Glucose level decreased from 100% to approx 20% basal level over 6hrs.
17	Stainless steel	–	260	80	800	296 (2 cm^2)	Sprague-Dawley saved	Pre-treatment, 2 min	Nanovesicles, 40 IU/ml, 3 h+ iontophoresis	Synergistic effect confirmed, preclinical/ clinical validation needed
43	Silicon (blunt)	$D_0$80 $D_1$25	150	–	–	100	rat	Hollow microneedles	Insulin	Not significant *in vivo*.
44	Stainless steel	–	200	50	1000	105	Sprague-Dawley hairless	Pre-treatment, 10s, 10 min and 4 h	Humulin-R, 100 IU/ml, 4 h on skin site	Shorter insert time, larger drops in glucose level, Inserting multiple times, smaller red. of glucose levels.
73	Nickel (10μ m wall thickness)	300 (base) 75 (top)	–	–	500	16	rat	Pre-treatment + passive diffusion 4 h	Insulin	After 30 min lag time, blood glucose level steadily reduced over 4 h, and remained constant over 4 h thereafter.
76	Silicon <100 mn	108	–	–	400	25 (21 hollow) 4 × 4 mm^2	Male Sprague-Dawley (hair removed)	Hollow microneedles	Lispro insulin (Humalog, 100 IU/ml)	Potentially useful for fast-acting insulin as a feasible option to improve glycaemic control.
78	Polymer 5–10μ m	300	–	–	500	100 (1 cm^{-})	Male Wistar rats	Biodegradable	3 min on skin, 0.37 IU up to 6.08 IU	Glucose level reacts upon insertion as expected. Clinically relevant studies needed.
79	as ref. (76)	108	–	–	400	as ref. (76)	as ref. (76)	Hollow microneedles	Humalog, 100 IU/ml	First active MEMS integrated infusion system for insulin (2 μl/h leak-free).

D_0: outer needle diameter, D_1: inner diameter or largest width of elliptical flow-through channel

The various tips depicted in Figure 8.6 all present design varieties that create a "mechanical cut" and may be defined by their main function: "cutting". However, their actual function is to release a controlled dose of insulin, and clinicians need to know how the exact size, type and geometric shape of the microconduits into the skin influence its delivery mechanism. When studying the pharmacodynamics of insulin release, one should not forget that the skin is an immunosensitive organ. When incisions are made in it, allowing foreign bodies to enter, an immune response may be triggered, which may alter the efficacy of insulin uptake, and also provoke unfavorable side-effects. This may also be an explanation for the smaller reduction in blood glucose levels observed by Martanto et al. when microneedles were inserted for a longer time or more than once [33]. Quality control of sterility, biocompatibility and efficacy of drug delivery are key for the final product.

Quantification of insertion and delivery processes, as well as the visualization of the incisions made by the individual types of microneedles in combination with the use of specific applicator devices, are therefore very important factors in these devices. A comparison on different types of insertion procedures and microneedle designs in transdermal drug delivery has been made by the Bouwstra group, but as yet not with respect to its effect on controlled insulin delivery [84]. The cost of statistically relevant clinical studies are high, and require a sustained supply of the devices which need approval by an ethical-medical committee. All the design varieties introduced and discussed above show a certain level of technological readiness that supports this request.

Clinical Perspective

From a clinical point of view, the microneedle array devices summarized schematically in Figure 8.6 are all of preliminary but potential interest, since the only studies that have used them have been small-scale animal trials ($n < 10$). Figure 8.9 shows two extreme design alternatives that have achieved a certain full proof-of-principle for insulin delivery, using out-of-plane microneedles as depicted in Figure 8.1(B) and (D). Figure 8.9(A) illustrates one such device integrated with a special actuator for the active release of insulin that is stored in a reservoir in liquid form. Figure 8.9(B) shows a simpler device, which releases insulin by a phase change, dissolving the polymeric material into the skin without any further active device components [78]. These self-dissolving microneedles encapsulate the insulin in a chondroitin sulfate matrix during micromolding of arrays, which is

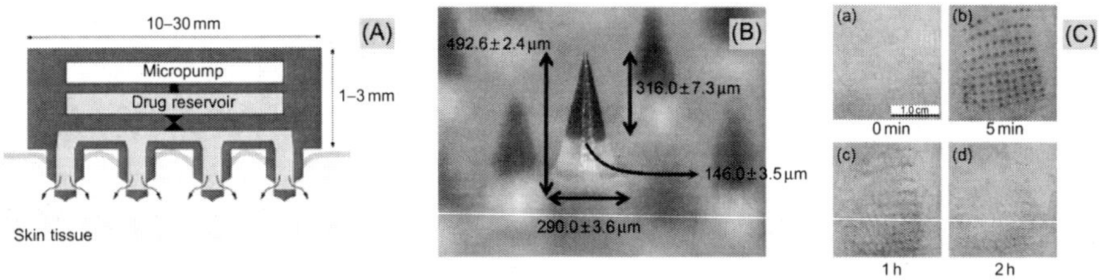

FIGURE 8.9

Illustration of design alternatives for insulin-releasing mechanisms. (A): integrated microneedle-micropump patch design. (Reproduced with permission from reference [43]. Copyright ©2008 Springer Science+Business Media, LLC.). (B): partially self-dissolving micropiles fabricated by micromolding in arrays from 10 x 10 microneedles. (C): skin prior to insertion (a), 5 min (b), 1 h (c) and 2 h (d) after release of insulin into rat skin. The dark color due to Evans blue staining disappeared and there was no damage on the rat skin. (Reproduced with permission from reference [78]. Copyright ©2009 Springer Science+Business Media, LLC.)

done by using flexible silicone (PDMS) molds. Both design alternatives demonstrate a viable solution for ultra-minimally invasive insulin release.

Computer-assisted design studies, in combination with the appropriate testing in a clinical laboratory, are needed to prove the microneedle concept. For diabetes care management, the devices introduced by Stemme's group offer stop-and-go delivery, and keep the device attached to the skin [32, 43, 76, 79], which may be of particular importance during food intake or exercise. It is not yet clear which type of device offers the best number and type of pathways through the skin. One technical solution for the investigation of the pharmacokinetics of insulin is to use a hollow microneedle array together with sensor augmented active pumping. The device introduced by Roxhed et al. currently shows the highest level of integration for such a class of drug delivery device [43]. Measuring glucose by a user-independent integrated sensor is also in development, as briefly mentioned in the previous section [70]. The augmentation of the released insulin dosage will require such feedback parameters, which can significantly vary with the patient's specific health status.

Any decision taken in the conceptual design phase of a new device will impact the type of experiments being carried out in drug delivery research. The choice of technology will therefore influence the data that can be collected for a clinical model. Insulin release by microneedle array pre-treatment has been taken to a fairly high level of testing, yet there are little or no data available that compare this novel transdermal mechanism of insulin release with the gold standard (bolus injection) within a clinically relevant number of human diabetic subjects. Although the working principle of insulin microneedle delivery has been confirmed by Gupta et al., utilizing a single, pulled glass needle at various insertion depths [49], there is still insufficient evidence to show that a delivery system equipped with an out-of-plane microneedle array can deliver insulin in a clinical application.

Delivery technology research can be carried out for the creation of a multitude of parallel injection sites with the preliminary devices presented here, but some level of concern remains regarding the degree of control that is exerted over the device insertion mechanisms. For example, applying high pressures during insertion will affect the skin properties, and thus the results of the transport studies. A delivery platform that includes sensor-augmented release is clearly desired, but at this stage of development many kinds of out-of-plane MNA devices can be already translated from the pure engineering sciences to research in drug delivery technology. Patient trials, especially those related to comfort, user-friendliness and pain sensation during the use of such devices, have by now demonstrated an appropriate level of technology readiness for clinical translational research. The initial studies by Gupta et al. [49], photographed the insertion sites to qualitatively evaluate the difference between injection depths. Much more indepth studies of the efficacy of MNA injections will be needed to prove their clinical effectiveness. For micromachined microneedles, new benchmark methods must be adopted to evaluate the clinical relevance of the different designs and materials. Both of the two concepts presented in Figure 8.9 provide a starting point for a further analysis of the specific design criteria for the application of MNAs in insulin delivery.

From the viewpoints of microengineering, drug delivery technology and clinical translation, the current microdevices show clearly that microsystems engineering presents certain benefits compared to those designs that use simple fine mechanical solutions or even mechanical assembly. A formal design approach can offer an exquisite tool for finding appropriate technological solutions for the development of transdermal insulin delivery, based on the aforementioned out-of-plane microneedle designs [85].

8.2 MNA-4-INSULIN: A BRIEF EVALUATION

In addition to developing MNA fabrication technology, it is also important to create a clinical focus for the use of microneedles in insulin delivery. This review gives an overview of the different types of microneedle technologies, which already refer to the proof-of-principle of microneedle insulin delivery. With respect to drug delivery research, devices from the three well-known categories of technology (silicon micromachining, replication and precision engineering) show enhanced transdermal delivery rates for insulin. Specific efficacy/bioavailability data has not yet been gathered, however. The devices show few side effects (including safety), and little discomfort during application, but other clear advantages over the hypodermic needle/syringe concept cannot yet be confirmed, and there is a lack of data on pharmacodynamics for systemic human insulin uptake. Nevertheless, there is no doubt that passive microneedle integrated skin patches could be an affordable solution for the painless intradermal delivery of insulin. In conclusion, a sufficiently high technology readiness of a variety of MNAs has been demonstrated. However, none of the systems discussed in the literature are ready for the market, mainly due to the lack of clinically confirmed reference data. Verification of the device quality and pharmacokinetics must be benchmarked first.

8.3 CONCLUSIONS

Is it too early or is it already too late to start a profitable business based on MNA technology? Based on this discussion of the application needs for insulin, it appears that there is not yet any significant business based on using microneedle arrays. Does this mean we just sit and wait for someone to take an initiative, and find a solution for the year 2030? Obviously, this is a rhetorical question. At the current stage of development it is not possible to judge which type of MNA will best deliver insulin via the skin. There are, of course, also other means of delivery, and other treatments for diabetes, but the delivery of insulin by transdermal means should be studied seriously. A great deal of clinical and business study is needed before further clinical decisions can be taken. Further, it is important to realize that high-tech business development benefits from the collaboration between different research disciplines at the cross-over point: the application. For drug delivery applications it is important that researchers from the pharmaceutical, drug delivery and the engineering sciences meet as early as possible to define a common aim to work on.

Having conducted this desk research, I must ask myself what research questions remain for a microfabrication expert in this field, and yes, I can think of many. Obviously, it is important to ask this question from the viewpoint of each discipline taking part in the work, and then combine them into one joint-call proposal for a focused technology development in this specific sector of the life sciences to provide a solution for the 2030 problem. This should be anticipated by the researchers in this field together with appropriate industry sectors as early as possible. Only this approach will make sure that the required results from the different disciplines can come together at the required pace of development. A world-wide science and technology funding structure has been created that provides research subsidies for public–private partnership.

Lessons learned from this case study should inspire students in the engineering sciences to start talking to their colleagues in the pharmaceutical and biomedical sciences and get those partnerships running. Of course, this discussion may also inspire managers of key technologies to adopt microfabrication principles in totally different areas to those introduced here. This chapter may serve as a starting point for a new business case. An ultra early technology adapter, similar to Unversity of Twente's spin-off MyLife Technologies, may allow many more real-world applications of micro- and nanofabrication technology.

References

[1] J. Hogan, Lab on a chip. A little goes a long way, *Nature* **442** (2006) 351–352.

[2] E.X. Vrouwe, R. Luttge, I. Vermes, A. Van Den Berg, Microchip capillary electrophoresis for point-of-care analysis of lithium, *Clin. Chem.* **53** (1) (2007) 117–123.

[3] H.J.G.E. Gardeniers, R. Luttge, E.J.W. Berenschot, M.J. de Boer, S.Y. Yeshurun, M. Hefetz, et al., Silicon micromachined hollow microneedles for transdermal liquid transport, *IEEE J. Microelectromech. Syst.* **12** (6) (2003) 855–862.

[4] P.C. Mills, S.E. Cross, Transdermal drug delivery: Basic principles for the veterinarian, *Vet. J.* **172** (2) (2006) 218–233.

[5] S. Godefroy, M. Peyre, N. Garcia, S. Muller, D. Sesardic, C.D. Partidos, Effect of skin barrier disruption on immune responses to topically applied cross-reacting material, crm197, of diphtheria toxin, *Infect. Immun.* **73** (8) (2005) 4803–4809.

[6] C.D. Partidos, S. Muller, Decision-making at the surface of the intact or barrier disrupted skin: Potential applications for vaccination or therapy, *Cell. Mol. Life Sci.* **62** (13) (2005) 1418–1424.

[7] D.T. O'Hagan, R. Rappuoli, Novel approaches to vaccine delivery, *Pharm. Res.* **21** (9) (2004) 1519–1530.

[8] B.W Barry, Novel mechanisms and devices to enable successful transdermal drug delivery, *Eur. J. Pharm. Sci.* **14** (2) (2001) 101–114.

[9] M. Kendall, Engineering of needle-free physical methods to target epidermal cells for DNA vaccination, *Vaccine* **24** (21) (2006) 4651–4656.

[10] F. Chabri, K. Bouris, T. Jones, D. Barrow, A. Hann, C. Allender, et al., Microfabricated silicon microneedles for nonviral cutaneous gene delivery, *Br. J. Dermatol.* **150** (5) (2004) 869–877.

[11] J. Birchall, S. Coulman, M. Pearton, C. Allender, K. Brain, A. Anstey, et al., Cutaneous DNA delivery and gene expression in *ex vivo* human skin explants via wet-etch microfabricated microneedles, *J. Drug Targeting* **13** (7) (2005) 415–421.

[12] S. Coulman, C. Allender, J. Birchall, Microneedles and other physical methods for overcoming the stratum corneum barrier for cutaneous gene therapy, *Crit. Rev. Ther. Drug Carr. Syst.* **23** (3) (2006) 205–258.

[13] M. Pearton, C. Allender, K. Brain, A. Anstey, C. Gateley, N. Wilke, et al., Gene delivery to the epidermal cells of human skin explants using microfabricated microneedles and hydrogel formulations, *Pharm. Res.* **25** (2) (2008) 407–416.

[14] Z. Ding, E. Van Riet, S. Romeijn, G.F.A. Kersten, W. Jiskoot, J.A. Bouwstra, Immune modulation by adjuvants combined with diphtheria toxoid administered topically in BALB/c mice after microneedle array pretreatment, *Pharm. Res.* **26** (7) (2009) 1635–1643.

[15] Z. Ding, F.J. Verbaan, M. Bivas-Benita, L. Bungener, A. Huckriede, D.J. van den Berg, et al., Microneedle arrays for the transcutaneous immunization of diphtheria and influenza in BALB/c mice, *J. Controlled Release* **136** (1) (2009) 71–78.

[16] S.A. Coulman, A. Anstey, C. Gateley, A. Morrissey, P. McLoughlin, C. Allender, et al., Microneedle mediated delivery of nanoparticles into human skin, *Int. J. Pharm.* **366** (1–2) (2009) 190–200.

[17] H. Chen, H. Zhu, J. Zheng, D. Mou, J. Wan, J. Zhang, et al., Iontophoresis-driven penetration of nanovesicles through microneedle-induced skin microchannels for enhancing transdermal delivery of insulin, *J. Controlled Release* **139** (1) (2009) 63–72.

[18] F.J. Verbaan, S.M. Bal, D.J. van den Berg, J.A. Dijksman, M. van Hecke, H. Verpoorten, et al., Improved piercing of microneedle arrays in dermatomed human skin by an impact insertion method. *J. Controlled Release* **128** (1) (2008) 80–88.

[19] S.A. Coulman, D. Barrow, A. Anstey, C. Gateley, A. Morrissey, N. Wilke, et al., Minimally invasive cutaneous delivery of macromolecules and plasmid DNA via microneedles. *Curr. Drug Deliv.* **3** (1) (2006) 66–75.

[20] D.A. La Van, D.M. Lynn, R. Langer, Moving smaller in drug discovery and delivery, *Nat. Rev. Drug Discovery* **1** (1) (2002) 77–84.

[21] A.C.R. Grayson, R.S. Shawgo, Y. Li, M.J. Cima, Electronic MEMS for triggered delivery, *Adv. Drug Delivery Rev.* **56** (2) (2004) 173–184.

[22] S.Z. Razzacki, P.K. Thwar, M. Yang, V.M. Ugaz, M.A. Burns, Integrated microsystems for controlled drug delivery, *Adv. Drug Delivery Rev.* **56** (2) (2004) 185–198.

[23] C.D. Partidos, Delivering vaccines into the skin without needles and syringes, *Expert Rev. Vaccines* **2** (6) (2003) 753–761.

[24] M.R. Prausnitz, Microneedles for transdermal drug delivery, *Adv. Drug Delivery Rev.* **56** (5) (2004) 581–587.

[25] Y.B. Schuetz, A. Naik, R.H. Guy, Y.N. Kalia, Emerging strategies for the transdermal delivery of peptide and protein drugs, *Expert Opin. Drug Deliv.* **2** (3) (2005) 533–548.

[26] S. Sharma, A.J. Nijdam, P.M. Sinha, R.J. Walczak, X. Liu, M.M.-C. Cheng, M. Ferrari, Controlled-release microchips, *Expert Opin. Drug Deliv.* **3** (3) (2006) 379–394.

[27] R.K. Sivamani, D. Liepmann, H.I. Maibach, Microneedles and transdermal applications, *Expert Opin. Drug Deliv.* **4** (1) (2007) 19–25.

[28] J. Vandervoort, A. Ludwig, Microneedles for transdermal drug delivery: A minireview, *Front. Biosci.* **13** (5) (2008) 1711–1715.

[29] A. Arora, M.R. Prausnitz, S. Mitragotri, Micro-scale devices for transdermal drug delivery, *Int. J. Pharmaceutics* **364** (2) (2008) 227–236.

[30] H. Kalluri, A.K. Banga, Microneedles and transdermal drug delivery, *J. Drug Deliv. Sci. Technol.* **19** (5) (2009) 303–310.

[31] J.D. Zahn, N.H. Talbot, D. Liepmann, A.P. Pisano, Microfabricated polysilicon microneedles for minimally invasive biomedical devices, *Biomed. Microdevices* **2** (4) (2000) 295–303.

[32] L. Nordquist, N. Roxhed, P. Griss, G. Stemme, Novel microneedle patches for active insulin delivery are efficient in maintaining glycaemic control: An initial comparison with subcutaneous administration, *Pharm. Res.* **24** (7) (2007) 1381–1388.

[33] W. Martanto, S.P. Davis, N.R. Holiday, J. Wang, H.S. Gill, M.R. Prausnitz, Transdermal delivery of insulin using microneedles *in vivo*, *Pharm. Res.* **21** (6) (2004) 947–952.

[34] B. Stoeber, D. Liepmann, Arrays of hollow out-of-plane microneedles for drug delivery, *J. Microelectromech. Syst.* **14** (3) (2005) 472–479.

[35] P. Griss, H.K. Tolvanen-Laakso, P. Merilinen, G. Stemme, Characterization of micromachined spiked biopotential electrodes, *IEEE Trans. Biomed. Eng.* **49** (6) (2002) 597–604.

[36] P. Griss, G. Stemme, Novel, side opened out-of-plane microneedles for microfluidic transdermal interfacing, in: *Proceedings of the IEEE Micro Electro Mechanical Systems (MEMS)* 2002, pp. 467–470.

[37] P. Griss, G. Stemme, Side-opened out-of-plane microneedles for microfluidic transdermal liquid transfer, *J. Microelectromech. Syst.* **12** (3) (2003) 296–301.

[38] J.G.E. Gardeniers, J.W. Berenschot, M.J. De Boer, Y. Yeshurun, M. Hefetz, R. Van't Oever, et al., Silicon micromachined hollow microneedles for transdermal liquid transfer, in: *Proceedings of the IEEE Micro Electro Mechanical Systems (MEMS)*, 2002, pp. 141–144.

[39] S. Henry, D.V. McAllister, M.G. Allen, M.R. Prausnitz, Microfabricated microneedles: A novel approach to transdermal drug delivery, *J. Pharm. Sci.* **87** (8) (1998) 922–925.

[40] D.V. McAllister, S. Kaushik, P.N. Patel, J.L. Mayberry, M.G. Allen, M.R. Prausnitz, Solid and hollow microneedles for transdermal protein delivery, *Proc. Controlled Release Soc.* (**26**) (1999) 192–193.

[41] D.V. McAllister, P.M. Wang, S.P. Davis, J.-H. Park, P.J. Canatella, M.G. Allen, et al., Microfabricated needles for transdermal delivery of macromolecules and nanoparticles:

Fabrication methods and transport studies, *Proc. Natl. Acad. Sci.* USA **100** (Suppl. 2) (2003) 13755–13760.
[42] F.J. Verbaan, S.M. Bal, D.J. van den Berg, W.H.H. Groenink, H. Verpoorten, R. Lttge, et al., Assembled microneedle arrays enhance the transport of compounds varying over a large range of molecular weight across human dermatomed skin. *J. Controlled Release* **117** (2) (2007) 238–245.
[43] N. Roxhed, P. Griss, G. Stemme, Membrane-sealed hollow microneedles and related administration schemes for transdermal drug delivery, *Biomed. Microdevices* **10** (2) (2008) 271–279.
[44] M.A. Ling Teo, C. Shearwood, K.C. Ng, J. Lu, S. Moochhala, *In vitro* and *in vivo* characterization of MEMS microneedles, *Biomed. Microdevices* **7** (1) (2005) 47–52.
[45] J.H. Park, M.G. Allen, M.R. Prausnitz, Biodegradable polymer microneedles: fabrication, mechanics and transdermal drug delivery, *J. Controlled Release* **104** (1) (2005) 51–66.
[46] M.I. Haq, E. Smith, D.N. John, M. Kalavala, C. Edwards, A. Anstey, et al., Clinical administration of microneedles: Skin puncture, pain and sensation, *Biomed. Microdevices* **11** (1) (2009) 35–47.
[47] A.L. Teo, C. Shearwood, K.C. Ng, J. Lu, S. Moochhala, Transdermal microneedles for drug delivery applications, *Mater. Sci. Eng.* **B132** (1–2) (2006) 151–154.
[48] G. Aggarwal, A. Garg, S. Dhawan, Transdermal drug delivery: Evolving technologies and expanding opportunities, *Indian J. Pharm. Educ. Res.* **43** (3) (2009) 251–259.
[49] J. Gupta, E.I. Felner, M.R. Prausnitz, Minimally invasive insulin delivery in subjects with type 1 diabetes using hollow microneedles, *Diabetes Technol. Ther.* **11** (6) (2009) 329–337.
[50] J.-H. Park, M.G. Allen, M.R. Prausnitz, Polymer microneedles for controlled-release drug delivery, *Pharm. Res.* **23** (5) (2006) 1008–1019.
[51] T. Matsuda, M. Mizutani, Liquid acrylate-endcapped biodegradable poly(-caprolactone-co-trimethylene carbonate). ii. Computer-aided stereolithographic microarchitectural surface photoconstructs, *J. Biomed. Mater. Res.* **62** (3) (2002) 395–403.
[52] R. Luttge, E.J.W. Berenschot, M.J. de Boer, D.M. Altpeter, E.X. Vrouwe, A. van den Berg, et al., Integrated lithographic molding for microneedle-based devices, *J. Microelectromech. Syst.* **16** (4) (2007) 872–884.
[53] G. Zhao, W. Li, G. Tang, J. Chen, Fabrication of bulk titanium out-of-plane microneedles, in: *4th IEEE International Conference on Nano/Micro Engineered and Molecular Systems, NEMS 2009*, 2009, pp. 428–431.
[54] A. Doraiswamy, C. Jin, R.J. Narayan, P. Mageswaran, P. Mente, R. Modi, et al., Two photon induced polymerization of organic-inorganic hybrid biomaterials for microstructured medical devices, *Acta Biomater.* **2** (3) (2006) 267–275.
[55] K. Lee, H.C. Lee, D.-S. Lee, H. Jung, Drawing lithography: Three-dimensional fabrication of an ultrahigh-aspect-ratio microneedle, *Adv. Mater.* **22** (4) (2010) 483–486.
[56] Q. Zhu, V.G. Zarnitsyn, L. Ye, Z. Wen, Y. Gao, L. Pan, et al., Immunization by vaccine-coated microneedle arrays protects against lethal influenza virus challenge, *Proc. Natl. Acad. Sci.* USA **106** (19) (2009) 7968–7973.

[57] B. Al-Qallaf, D.B. Das, Optimization of square microneedle arrays for increasing drug permeability in skin, *Chem. Eng. Sci.* **63** (9) (2008) 2523–2535.
[58] Y.-G. Lv, J. Liu, Y.-H. Gao, B. Xu, Modeling of transdermal drug delivery with a microneedle array, *J. Micromech. Microeng.* **16** (11) (2006) 2492–2501.
[59] A.K. Banga, Microporation applications for enhancing drug delivery, *Expert Opin. Drug Deliv.* **6** (4) (2009) 343–354.
[60] H.S. Gill, M.R. Prausnitz, Pocketed microneedles for drug delivery to the skin, *J. Phys. Chem. Solids* **69** (5–6) (2008) 1537–1541.
[61] J. Ji, F.E.H. Tay, J. Miao, Microfabricated hollow microneedle array using ICP etcher, *J. Phys. Conf. Ser.* **34** (1) (2006) 1132–1136.
[62] S.J. Moon, C.Y. Jin, S.S. Lee, A novel method of microneedle array fabrication using inclined deep x-ray exposure, *J. Phys. Conf. Ser.* **34** (1) (2006) 180–186.
[63] R.F. Donnelly, D.I.J. Morrow, P.A. McCarron, A. David Woolfson, A. Morrissey, P. Juzenas, et al., Microneedle arrays permit enhanced intradermal delivery of a preformed photosensitizer, *Photochem. Photobiol.* **85** (1) (2009) 195–204.
[64] C.S. Kolli, A.K. Banga, Characterization of solid maltose microneedles and their use for transdermal delivery, *Pharm. Res.* **25** (1) (2008) 104–113.
[65] G. Li, A. Badkar, S. Nema, C.S. Kolli, A.K. Banga, *In vitro* transdermal delivery of therapeutic antibodies using maltose microneedles. *Int. J. Pharm.* **368** (1–2) (2009) 109–115.
[66] V.M. Meidan, B.B. Michniak, Emerging technologies in transdermal therapeutics. *Am. J. Ther.* **11** (4) (2004) 312–316.
[67] D.V. McAllister, M.G. Allen, M.R. Prausnitz, Microfabricated microneedles for gene and drug delivery, *Annu. Rev. Biomed. Eng.* **2** (2000) 289–313.
[68] D. Trebotich, J.D. Zahn, D. Liepmann, Complex fluid dynamics in bioMEMS devices: Modeling of microfabricated microneedles, in: *2002 International Conference on Computational Nanoscience and Nanotechnology - ICCN 2002*, 2002, pp. 20–23.
[69] D. Trebotich, J.D. Zahn, B. Prabhakarpandian, D. Liepmann, Modeling of microfabricated microneedles for minimally invasive drug delivery, sampling and analysis, *Biomed. Microdevices* **5** (3) (2003) 245–251.
[70] J.D. Zahn, Y.-C. Hsieh, M. Yang, Components of an integrated microfluidic device for continuous glucose monitoring with responsive insulin delivery, *Diabetes Technol. Ther.* **7** (2005) 536–545.
[71] M.R. Prausnitz, Overcoming skin's barrier: The search for effective and user-friendly drug delivery, *Diabetes Technol. Ther.* **3** (2) (2001) 233–236.
[72] P. Zhang, G.A. Jullien, Microneedle array for drug delivery and fluid extraction, in: *Proceedings, (2005) International Conference on MEMS, NANO and Smart Systems*, ICMENS Digital Object Identifier: 10.1109/ICMENS.2005.71.
[73] R. Zhang, P. Zhang, C. Dalton, G.A. Jullien, Modeling of drug delivery into tissues with a microneedle array using mixture theory, *Biomech. Model. Mechanobiol.* (2009) 1–10.
[74] S.P. Davis, W. Martanto, M.G. Allen, M.R. Prausnitz, Hollow metal microneedles for insulin delivery to diabetic rats, *IEEE Trans. Biomed. Eng.* **52** (5) (2005) 909–915.
[75] B. Ma, S. Liu, Z. Gan, G. Liu, X. Cai, H. Zhang, et al., A PZT insulin pump integrated with a silicon microneedle array for transdermal drug delivery, *Microfluid. Nanofluid.* **2** (5) (2006) 417–423.

[76] N. Roxhed, B. Samel, L. Nordquist, P. Griss, G. Stemme, Compact, seamless integration of active dosing and actuation with microneedles for transdermal drug delivery, in: *Proceedings of the IEEE International Conference on Micro Electro Mechanical Systems (MEMS)*, vol. 2006, 2006, pp. 414–417.

[77] Y. Ito, E. Hagiwara, A. Saeki, N. Sugioka, K. Takada, Feasibility of microneedles for percutaneous absorption of insulin, *Eur. J. Pharm. Sci.* **29** (2006) 82–88.

[78] Y. Ito, T. Yamazaki, N. Sugioka, K. Takada, Self-dissolving micropile array tips for percutaneous administration of insulin, *J. Mater. Sci. Mater. Med.* **21** (2) (2009) 835–841.

[79] N. Roxhed, B. Samel, L. Nordquist, P. Griss, G. Stemme, Painless drug delivery through microneedle-based transdermal patches featuring active infusion, *IEEE Trans. Biomed. Eng.* **55** (3) (2008) 1063–1071.

[80] A. Davidson, B. Al-Qallaf, D.B. Das, Transdermal drug delivery by coated microneedles: Geometry effects on effective skin thickness and drug permeability, *Chem. Eng. Res. Des.* **86** (11) (2008) 1196–1206.

[81] B. Al-Qallaf, D.B. Das, Optimizing microneedle arrays for transdermal drug delivery: Extension to non-square distribution of microneedles, *J. Drug Targeting* **17** (2) (2009) 108–122.

[82] M. Cormier, B. Johnson, M. Ameri, K. Nyam, L. Libiran, D.D. Zhang, P. Daddona, Transdermal delivery of desmopressin using a coated microneedle array patch system, *J. Controlled Release* **97** (3) (2004) 503–511.

[83] D.P. Wermeling, S.L. Banks, D.A. Hudson, H.S. Gill, J. Gupta, M.R. Prausnitz, et al., Microneedles permit transdermal delivery of a skin-impermeant medication to humans, *Proc. Natl. Acad. Sci.* USA **105** (6) (2008) 2058–2063.

[84] S.M. Bal, J. Caussin, S. Pavel, J.A. Bouwstra, *In vivo* assessment of safety of microneedle arrays in human skin, *Eur. J. Pharm. Sci.* **35** (3) (2008) 193–202.

[85] K.N. Otto, E.K. Antonsson, Trade-off strategies in engineering design, *Res. Eng. Des.* **3** (2) (1991) 87–103.

CHAPTER

Reflective Comments and Conclusions

9

CHAPTER CONTENTS

Many examples of micro-nanofabrication technology are already present all around us but remain unnoticed (or better, are in disguise). In this field of high-tech developments we keep asking when is it time to take a new technology at the nanoscale to industry? Personally, I am convinced that the micro- and nanoworld will penetrate into all areas of application, within technical and biological systems eventually. Like all other engineering processes, their exploitation will bring benefits, however, they may also present us with novel hazards. Scientists must make both aspects of their developments clear to the public, i.e., both the potential utility and potential negative effects. Ideally, we make sure that all involved risks are evaluated before the introduction of a novel technology. Modern society considers the estimation of risk levels in economical, environmental and medical context to be an overall responsibility of government and industry alike.

We have already learned a great deal about risk management from the programs for the semiconductor/microelectronics industry. Businesses have procedures in place that monitor (often also thanks to affordable (miniaturized) sensor technology) working environments which contain hazardous chemicals and machinery, can evacuate their facilities if unpredictable conditions occur, and take all necessary actions to check on the safety of the infrastructure before resuming operations. Miniaturization and the use of nanomaterials in industry may create other, as yet unknown risks, which will have to be critically evaluated against the overall benefits of greener

Microfabrication for Industrial Applications. DOI: 10.1016/B978-0-8155-1582-1.00009-5

chemistry, novel nanomedicines and safe-guarding sensor technology that are enabled by micro-nanofabrication. I would like to discuss the environmental aspects of nanotechnology, and specifically address the health aspects of very small artificially created particles in this final chapter.

9.1 ENVIRONMENTAL ASPECTS

All manufacturing industries must consider the possible negative effects of the technology they use, both in the product development phase and in the product itself. These concerns start from the refining of the raw materials, and continue to the creation of semi-finished products with specific material properties, as well as other related processes that have a possible impact on the environment throughout the product's life-cycle. It is difficult to estimate the environmental impact of a micro-nanofabricated product's full life-cycle and we will not try to generalize or condense this topic into one single section of a book chapter.

Nevertheless, I would like to address this topic, at least briefly, because of the urgent need to incorporate such considerations into micro-nanofabrication product design. Many dedicated funding schemes worldwide, and this book are intended to further stimulate the implementation of micro-nanofabrication into industrial application, and this should not be hampered simply by the lack of awareness.

Similarly to the previous chapters, this broad topic of environmental aspects of micro-nanofabrication is best addressed by an example from industry. I found this example by Williams et al. in the field of electricity-generating technologies who question whether photovoltaic (PV) systems, although producing little or no environmental pollution at the point of use, carry a significant environmental burden because of the types of materials and fabrication methods involved in their production [1]. Without going into their life-cycle analysis, we know that current state of the art is clearly capable of producing PV systems at a very large scale. Williams et al. conclude that there are significant opportunities for increasing the environmental performance of a PV system. This is where micro-nanofabrication may be able to offer technological solutions, e.g., by replacing currently-used batteries by novel means of energy storage systems.

A discussion of the possible routes for higher efficiency in energy storage systems is beyond the scope of this book. We can, of course, conduct desk research and bring all the different factors of currently existing scientific strategies to the point. Just to give an idea of novel concepts being developed in this field Birch examines the construction of an *artificial leaf* by means of micro-nanofabrication [2]. This example demonstrates that a micro- and nanofabrication concept may improve the overall efficiency of a given amount of available resources, and by itself does not form a direct threat to the environment, but the processes and materials in the product's life-cycle still might.

It is therefore possible that micro- and nanotechnology could be used to optimize systems, by stimulating the selection of materials for a cradle-to-grave approach, to strongly reduce or even deminish completely an environmental negative impact of novel commercial products and processes [3–6].

9.2 HEALTH ASPECTS OF NANOPARTICLES

One of the most alarming examples of health risks we have experienced in recent years arising from technical solutions is the case of asbestos in the building industry. Asbestos was the number one material for fire protection and fire resistance. The material as used appears harmless, but exposes its danger upon aging, or when machined by drilling, etc. Under these circumstances fine-scale particles of a few micrometers in length are released. These whisker-type structures get stuck in the cellular tissue, specifically in the lung, initially causing a reduction in lung capacity, and also a possible carcinogenic reaction. Currently, these symptoms have no known cure.

This material has now been banned from production and use, and significant effort is made by industry and government to remove it from existing buildings. Although asbestos is now going to the grave to minimize future health risks, we have to realize that industries and governments have the responsibility of investigating possible health risks from the introduction of novel materials and products before they go to market.

Nanoparticles (μNP) are starting to be used in commercial products. These particulate systems are either designed as a stand-alone nanodevice in, e.g., drug delivery applications, or are used as the smallest building block within higher order architectures, e.g., in their incorporation into self-healing materials [7, 8]. These materials do not constitute a new risk to mankind. μNPs are all around us in nature, and many natural and artificially designed materials already contain them.

Our novel understanding of these μNP systems allows their controlled placement in artificially created systems at the molecular or atomic scale, e.g., from solutions or the gas phase. If the μNPs remain either in solution or within the solids created during a process, the materials are considered as safe. This concept has yet to be evaluated over the product's full life-cycle. Of course, in cases that use powderized μNPs, special safety rules apply and are stringently followed, since we have learned so much about asbestos and its related diseases.

However, I believe that we are quite capable of organizing and installing appropriate rules for dealing with these substances. Like any other chemical, the health risks of μNPs can be identified and communicated through the appropriate channels within working environments.

Although I am not an expert on the design of such particulate systems, I do work on microreaction controlling technologies on chip that enable their fabrication, so I feel that it is important to mention these facts within the content of this book. I wish to point out that the risk is not that associated with working with μNPs. The real risk arises from combining artificially produced materials, by trial and error, and only considering the utility of the outcome. Scientists working in micro-nanofabrication have the opportunity to accumulate appropriate knowledge to counteractary any possible negative effects and any associated predictable risks when the materials are affiliated with traceable drivers. Scientific publication will not be enough; the results of research findings in μNPs should be disseminated via channels used by the general public. As a first source on μNP fabrication and their use I refer the reader to the contributions by Rottelo and co-workers [9] and Yang et al. [10].

In sum, I would like to suggest putting all the different types of effort concerning μNPs into one public–private research program, to investigate the design, ethical and technical validation of these novel technologies, as well as their clinical and economical impacts on society. Fragmented investigation approaches may simply spend tax-payers' money on similar debates to those experienced in the development of genetically-modified food and organic feedstock. Such a process should not only provide a solution to a single technical problem (e.g., treating a specific disease) but should also provide long-term perspectives on the safety and sustainability of μNP technology.

9.3 CONCLUSIONS

This book is about microfabrication for industrial applications. It provides a first insight to the variety of technical processes and initial efforts regarding the implementation of these novel technologies in microdevices, e.g., sensors and actuators. Of course, the fabrication technologies at the nanoscale are part of this field, and become more and more important in applications. Utilizing micro- and nanofabrication technologies in real-world tangible products (raw materials, semi-finished or fully developed end-user products) is a task for the applied sciences and industries. These actors meet in the microsystems devices industry, which has evolved from the previously established and highly successful semiconductor industry.

Many recent innovations have been enabled by microfabrication. In this book we have studied a variety of cases at different levels of development. Resonant microsensors, for example, are already highly established in the market (accelerometers in our cars and mobile phones). The introduction of technologies including *wet-state* components is less advanced. Although microfluidics is being investigated by different scientific disciplines, its application in commercial products is still hampered by the complexity of systems design. On the other hand, once a viable outlet has been identified, e.g., working towards a point-of-care diagnostic device dedicated to the measurement of lithium in blood (see the case in Chapter 7), these technologies are clearly able to penetrate into their market place. Cross-disciplinary efforts produce rapid progress in the field of nanotechnologies. Nanoscience and nanotechnology anticipate also on the mimicking of many processes in biology that occur naturally at the nanoscale, and by the introduction of this novel field of work, societies can reach far beyond microtechnology as an occupation of engineering scientists alone.

However, there has not been a paradigm shift that would be comparable to the effect of the fabrication of the integrated electronic circuit according to Moore's law[1] on the world's overall living standards. These devices are reaching the physical limit of a line width that can act as a transistor, which signals the end of the microelectronic development era. This is why novel nanofabrication techniques must emerge to provide other means of fulfilling this logic function within the design of future systems. One concept is to carry out logical operations at the molecular level, switching, for example, the energy state of a π-bond in an organic chemical compound. The ability to create materials and systems from the bottom up is exciting and may indeed present society with a new paradigm shift.

[1]Moore's law refers to the doubling of the number of devices within the integrated chip by reduction of the gate-oxide width in the design of the transistor by innovation in lithography every 18 months.

These efforts will eventually provide us with solutions for specific needs. The nanosciences and nanotechnologies will feed the knowledge grid as much as other areas of knowledge-creating disciplines, and will create new disciplines, which eventually will lead to the continuous generation of marketable non-tangible and tangible products in a sustained environment. In the latter, microfabrication will remain a stable factor in the overlapping networks of industrial and academic actors that now, more than ever, explore the working mechanisms of smaller and smaller dimensions, in which MST offers more than Moore.

References

[1] T. Williams, J. Guice, J. Coyle, Strengthening the environmental case for photovoltaics: A life-cycle analysis, in: *Conference Record of the 2006 IEEE 4th World Conference on Photovoltaic Energy Conversion, WCPEC-4*, vol. **2**, 2007, pp. 2509–2512.

[2] H. Birch, The artificial leaf using sunlight to split water molecules and form hydrogen fuel is one of the most promising tactics for kicking our carbon habit. Hayley Birch examines the options, *Chem. World* **6** (5) (2009) 42–45.

[3] P. Ellis, K.C. Desouza, On information management, environmental sustainability, and cradle to cradle mentalities: A relationship framework, *Bus. Inf. Rev.* **26** (4) (2009) 257–264.

[4] J. Potting, C. Kroezey, Cradle to cradle: Old wine or new spirits, *Integr. Environ. Assess. Manage.* **6** (2) (2010) 316–317.

[5] C.A. Bakker, R. Wever, C. Teoh, S. de Clercq, Designing cradle-to-cradle products: A reality check, *Int. J. Sustainable Eng.* **3** (1) (2010) 2–8.

[6] H. De Ridder, R. Beheshti, Innovative, dynamic and collaborative technologies for integral design, *Int. J. Des. Sci. Technol.* **16** (2) (2009) 121–134.

[7] M.D. Hager, P. Greil, C. Leyens, S. Van Der Zwaag, U.S. Schubert, Self-healing materials, *Adv. Mater.* **22** (47) (2010) 5424–5430.

[8] Y. Jinglei, M.W. Keller, J.S. Moore, S.R. White, N.R. Sottos, Microencapsulation of isocyanates for self-healing polymers, *Macromolecules* **41** (24) (2008) 9650–9655.

[9] M. De, P.S. Ghosh, V.M. Rotello, Applications of nanoparticles in biology, *Adv. Mater.* **20** (22) (2008) 4225–4241.

[10] X.-C. Yang, B. Samanta, S.S. Agasti, Y. Jeong, Z.-J. Zhu, S. Rana, et al., Drug delivery using nanoparticle-stabilized nanocapsules, *Angew. Chem. Int. Ed.* **50** (2) (2011) 477–481.

Index

T